BEYOND THE BIG RUN

"I'll be lucky to last a bloody year here." Nineteen-year-old Charlie Schultz was not impressed when he first arrived at Humbert River station in 1928. Humbert River was then a tiny, run-down station in a tangle of wild ranges neighbouring the legendary Victoria River Downs in the Northern Territory. This was one of Australia's last frontiers. There were hostile blacks in the ranges, fences were almost unknown and wild cattle overran the stations. Transport was by horseback, donkey, camel or by foot and living conditions were primitive. There were no radios, telephones or flying doctors. Yet Charlie did last the year, and he went on to work the station for another forty-three years.

Beyond the Big Run tells the story of Charlie's struggle to turn Humbert into a first-class cattle station in the face of physical and financial hardship, loneliness, and the wild bush itself. His stories are set against the colourful characters and events of the Victoria River district — the stockmen and station managers, horse thieves and police, gun fights and spearings, Christmas sports and horse races — a society and a way of life that is now gone forever.

Cover design by Michael Ward, featuring a portrait of Charlie Schultz at twenty-two years of age against a photograph of Humbert River cattle watering at the Murranji waterhole, 1935.

Charlie Schultz was born in Charters Towers, Queensland, in 1908. At nineteen years of age, Charlie was thrust into the role of owner and manager of Humbert River Station in the Victoria River district. In 1941, after years of struggle and loneliness, and a lot of hard work, Charlie was able to clear the debt on the property. In that year also, he married Hessie Graham and together they worked for the next three decades to create one of the showpiece stations in the Northern Territory. Hessie Schultz died in 1979, and Charlie now lives in Queensland.

Darrell Lewis first met Charlie Schultz briefly in 1981, but it was not until 1990, when he spent two and a half days listening spellbound as Charlie told him stories of events and characters in the Victoria River district, that the idea for the book took seed. Darrell Lewis was born in 1949 in Wagga Wagga, New South Wales. He left school at sixteen and spent many years working in a variety of jobs in remote outback areas. In 1978 he enrolled at the Australian National University. By 1990 he had completed a masters degree in archaeology and written a book on Arnhem Land rock paintings. Since 1990 he has worked primarily on historic site surveys in the Victoria River district.

BEYOND THE
BIG RUN

STATION LIFE IN
AUSTRALIA'S LAST FRONTIER

Charlie Schultz
Darrell Lewis

University of Queensland Press

First published 1995 by University of Queensland Press
Box 42, St Lucia, Queensland 4067 Australia
Reprinted 1995

© Darrell Lewis 1995

This book is copyright. Apart from any fair dealing
for the purposes of private study, research, criticism
or review, as permitted under the Copyright Act, no
part may be reproduced by any process without written
permission. Enquiries should be made to the publisher.

Typeset by University of Queensland Press
Printed in Australia by McPherson's Printing Group

Distributed in the USA and Canada by
International Specialized Book Services, Inc.,
5804 N.E. Hassalo Street, Portland, Oregon 97213-3640

Sponsored by the Queensland Office
of Arts and Cultural Development.

Cataloguing in Publication Data
National Library of Australia

Schultz, Charles, 1908–
  Beyond the big run.

  Bibliography.
  Includes index.

  1. Schultz, Charles, 1908– . 2. Ranchers — Northern
  Territory — Victoria River Region — Biography. 3. Ranch life
  — Northern Territory — Victoria River Region — Biography.
  4. Victoria River Region (N.T.) — History. I. Lewis, D.
  (Darrell). II. Title.

636.213092

ISBN 0 7022 2650 5

Contents

Preface *vii*

1 Early Years *1*
2 Buying Humbert River Station *17*
3 "The Great Australian Loneliness" *28*
4 Outlaws *42*
5 The Big Run *62*
6 VRD Outstations *76*
7 The Upper Wickham *93*
8 To Queensland, Droving *105*
9 A Lot To Be Thankful For *126*
10 Wife and Kids *137*
11 The End of Isolation *152*
12 Rough Hewn Men *170*
13 They Thought I'd Gone Mad *183*
14 Bullita and Whitewater *196*
15 Beyond Humbert *215*

Notes *224*
Key Events *229*
Glossary *231*
Select Bibliography *233*
Index *236*

Acknowledgments

As always, I owe the greatest acknowledgment to my wife, Debbie Rose. As well as providing financial support while the book was being written, Debbie read transcripts, offered invaluable advice on the structure of the book, corrected various infelicities, and put up with a sometimes obsessive author. Many other people have contributed to this book. Dr Peter Read of Canberra and Donna Cattanach (nee Schultz) of Hindmarsh Island read and commented on early drafts of the manuscript. Nola Lewis and Samantha Wells both helped correct the proofs.

Most of the photographs used in the book come from the personal collection of Charlie Schultz. Sources for other photographs include the Roden, Walker, Schultz (C.E.) and Mahoney Collections (Northern Territory Archives), and the Feast Collection (National Museum of Australia). My special thanks go to Bobby Buchanan (Darwin) for her constant support and the use of photographic and manuscript material, and to Dave Magoffin (Brisbane), John Gordon (Mundaring), Marie Mahood (Cattle Camp Station, Qld) Bert Mettam (Adelaide), "Darky" Pollard (Ravenswood) and I.J. Raymond (NSW), all of whom allowed the use of photographs from their private collections.

Finally, I gratefully acknowledge the financial assistance received from the Northern Territory Government via the NT History Awards.

Darrell Lewis

Imperial — Metric Conversion

Australian currency was based on the pound (£) which was converted to $2.00 in 1966. The pound consisted of 20 shillings (20/-) of 12 pence (12d) each. The pound was known colloquially as a "quid" and the shilling as a "bob".

One inch	=	2.5 centimetres
One foot (12 inches)	=	30.4 centimetres
One yard (3 feet)	=	91.4 centimetres
One mile (1760 yards)	=	1.6 metres
One square mile	=	2.6 square kilometres
One acre	=	0.4 hectares
One gallon (8 pints)	=	4.55 litres
One pound (16 ounces)	=	0.45 kilograms
One hundredweight (112 pounds)	=	50.8 kilograms
One ton (20 hundredweight)	=	1016 kilograms

Preface

This book is the product of two approaches to Victoria River history: Charlie Schultz lived it for forty-four years, mustering the wild cattle, creating a first-class station from virgin bush, experiencing many important events firsthand and absorbing the campfire yarns of men — black and white — many of whom were "old-timers" when he first went to Humbert River station.

I arrived in the Victoria River country the year Charlie left it — 1971 — and soon became interested in the region's history. Over the next two decades I worked in the region in various capacities and took every opportunity to visit remote "back-country" areas and historic sites. In 1980 I began intensive archival research, seeking out the diaries, memoirs, police reports and photographs that document the little-known history of the district.

Charlie and I have thus produced this book through our shared knowledge and love of the Victoria River country — the vast plains, wild ranges and big rivers — and through our shared knowledge of the rugged individuals and events that constitute Victoria River history.

In putting together Charlie's stories I have not sought to "correct" every difference between his recollections and contemporary accounts. Where there was an indisputable error, for example, where Charlie thought a particular station manager was involved in an event and the documents showed that that person had already left the district, I discussed the anomaly with Charlie, and corrected it accordingly.

In most instances, however, I left the "facts" of Charlie's stories as he told them, irrespective of differences from the contemporary record. After all, contemporary records are subject to errors of memory, or, for a variety of reasons, deliberate alteration or censorship. In any case, the power of Charlie's stories is not dependent on the historical accuracy *vis-a-vis* historic documents, although there is remarkable congruence between the two sources. The real power is in the telling, and in the fact that for Charlie, much of what he had to say is not only history — it is his life.

I have not tried to transform Charlie's accounts into standard English. As far as possible, I have retained the colour and style of Charlie's speech — the idiom of a Northern Territory bushman. All of the stories were tape-recorded and transcribed (verbatim). I then "cleaned up" the transcriptions, removing the "umms" and "ahhs" and deleting repetitious words. Some of the information was reorganised to facilitate the flow of the story, and finally I organised the stories into chapters.

A few of the terms Charlie uses may be regarded by some as offensive. However, in the context of his times (and in the manner in which he now uses these expressions), they were normal, acceptable words. Readers would do well to look beyond the language to the man, a man who was compassionate, humane, innovative, and in many ways ahead of his time. A man described by various of his contemporaries as "one of the best horsemen in the Territory", "one of the best cattlemen in the Territory", "one of the toughest men in the Territory", "the gentleman drover", "a half a Ned Kelly or something", and "one man in about ten thousand".

Darrell Lewis
Darwin

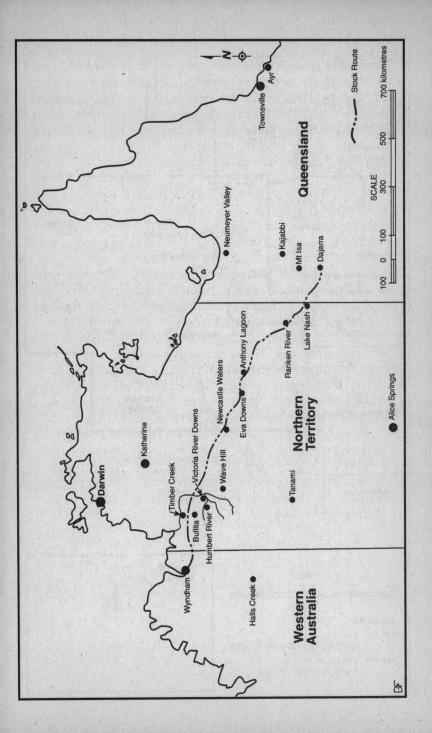

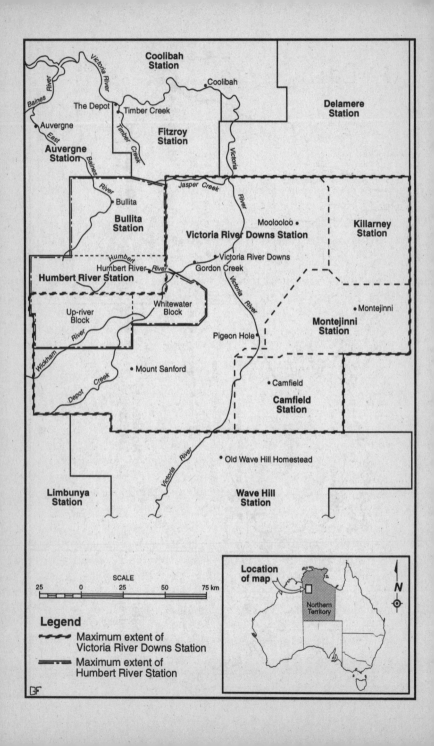

1 Early Years

> For good undone and gifts misspent and resolutions vain,
> 'Tis somewhat late to trouble. This I know —
> I should live the same life over, if I had to live again;
> And the chances are I go where most men go.
>
> 'The Sick Stockrider', Adam Lindsay Gordon, 1870

There was one night that nearly finished me with Humbert. I'd been on the place for about two years, the last twelve months alone. I had my bunk inside the house between two doorways and about three o'clock in the morning — pitch dark — I heard the dogs barking. I was always a very light sleeper those days and I woke. I heard one dog bark and then another growling. A half-grown cattle dog was chained up right near my door and I heard him bark when the other dogs weren't barking. I thought, "That's funny?" and sat up in my bed. The young dog barked again.

There were still wild blacks in the bush then, so I always camped with a revolver under my pillow and a rifle alongside of me — always loaded of course. I scrambled out from under my bush net and grabbed the rifle. I had the sense to stand to one side instead of in front of both doors, so no one could pick me off. I knew they couldn't do anything from the sides because the walls were solid timber.

I sang out: "Who's there?" No answer. I waited a while and the dog started growling again. I could hear him moving on his chain. "That bugger can see something and it's not likely to be a dingo," I thought. I cocked the rifle. I stood right near the door then in case anyone was going to come through. A thousand thoughts flashed through my mind: "Is a mob of blacks going to rush me? Will I shoot on sight?" If it was blacks I knew they couldn't give much trouble outside, but they could at the doors, so I just stood there with the rifle cocked and I thought, "Well I'll get the first bloke anyway."

I called out: "If you don't answer I'll shoot!" Still no answer. I was getting more stirred up all the bloody time, and next minute the dog sat back on the chain as though somebody was trying to wallop him with a stick. With that, I jumped outside quick smart with the rifle ready. I was looking a dozen places at once, trying to see who or what was there, but about all I could see in the dark was the bloody dog sitting back on the chain. Then the dog started yelping like hell and the next bloody minute something hit me on the leg. Oh God — I nearly went through the sky!

It was a bloody chook! I had some fowls that used to roost in a tree alongside the house, and one fell out of the tree apparently. They get blind at night and it blundered down towards the house. It was heading up to this half-grown cattle dog and the dog was taking ever-increasing fright. And then it hit me on the leg. By gees, I tell you what, that gave me a bloody fright. I went inside and I sat on my bloody bed, and I thought, *"Is it worth it?"* I'd had it. I was twenty or twenty-one and I cried that night — I'd *really* had it. I thought, "Bugger this, life's not worth it. I'm gettin' out of this place tomorrow. I'll ride over to VRD and I'll leave my bloody horses there and I'll get out somehow."

I didn't go to sleep again, but by the time daylight came I'd calmed down a bit. I heard pots rattling — gins putting on the billycans — and then I heard one of them sing out: "Billycan bin jump up, boss." I got out of bed and went over. I forget what I did again that day — I probably went for a ride, but by Christ I got a fright that night. If I'd seen a human being I would've shot them.

That's an indication of what it was like for me on Humbert

in the early years — the first seven years were the hardest. When I first went out there in 1928, Humbert River Station was just a few bark huts. No one had lived there for about five months and the place was frightfully run down: saddles were falling to pieces, there were horses that should have been broken in and unbranded cattle everywhere. There was only one half-rotten yard and on top of all this the place was £8000 in debt. I took one look at it all and thought, "I'll be lucky to last a bloody year here." I left Humbert forty-four years later.

To explain just how I came to be on Humbert in the first place, I'll need to go back to when I was a kid. In fact, I might as well go back to the beginning — what little I know about it. I don't know where either of my maternal grandparents came from. My mother's father, Edward Larkin, married Kate Hoolihan and they raised a family of nine children, five boys and four girls. My mother, Bridgette Agnes, was the third eldest child.

Early on Ted Larkin managed a mine at Charters Towers and later he bought a bush pub at Clare, on the Burdekin River. For a time this was a good location for a pub because the Burdekin River crossing was on the mail coach route between Bowen and Reed River. It also was on a major stock route that carried thousands of cattle into newly opened country to the north and west. Later the railway along the coast reduced traffic through Clare, so Ted sold out and settled permanently on a cane farm near Ayr.

My paternal grandfather, Wilhelm Schultz, came out from Germany on a sailing ship when he was nineteen. My grandmother-to-be, Alberta Bartz, travelled out on the same boat as he did. She was only nine. They landed in Australia at a port just to the south of Sydney, and they caught up with each other about ten years later and married. They ended up in Queensland, because my father was born in Dalby in 1872. He was the eldest of nine children — six sons and three daughters — and they all grew up in that district.

I don't know much about Dad's childhood. He left school when he was about twelve or thirteen and got a job as a horse-tailer with some drovers, several brothers by the name of Fletcher. They lifted over a thousand head of big bullocks from Carandotta Station in western Queensland, and started for Victoria. Dad was getting paid a pound a week which doesn't sound like much now, but as he often said, for a pound you could buy a shirt and trousers, and also a pair of elastic-side boots; he reckoned people were just as well off those days as they were later on when wages were higher. From the time they left Carandotta they were in rain most of the way, but Dad reckoned no creeks or rivers would stop the Fletcher brothers. He was only young then, but he reckoned he could swim like hell. They'd give him a light rope to carry in his mouth and he'd swim across the rivers. The other end of that rope was tied to a heavier one, and once he got across to the other side he'd pull the heavier rope over. That was used to help pull the wagon across. They'd lay a big fly in the water, roll their wagon onto it and take off the wheels. Then they'd tie the sides of the fly up around the wagon, bail out the water, and make a kind of raft. Someone would get in front and other fellas would push behind, and they'd swim and pull it across.

It took them eleven months to do the trip and when the cattle were sold they only covered the cost of droving them down. Dad was paid off and went home to Dalby. The following year the Fletcher brothers asked him to come on another droving trip, but he told them he'd given droving away for a while. Later he regretted that he didn't go with them. The Fletcher brothers used to help some of their regular employees to get established on small cattle properties.

Well, eventually Dad bought a few horses and followed the Queensland coast up. I suppose he'd heard all about the stations that were opening up in north Queensland.

When he was about thirty or thirty-one he was droving a mob of horses to Lochinvar Station to meet one of his brothers there — Albert Schultz I think it was. They were heading out to go brumby running in the Gulf of Carpentaria country. I think they were after Timor ponies.

In those days a man named A.H.W. Cunningham owned Lochinvar, Woodhouse and Strathmore Stations. He ran against my father camped on the bank of the Burdekin River at a place known as "The Rocks", and asked him if he'd like a job.

> Old Edward Cunningham, now at Woodhouse Station on the Lower Burdekin, saw his twelve thousand good Shorthorns reduced to eighteen hundred. A government official sent in 1899 to make a conservative estimate of stock losses took tallies from many station owners, and reported that cattle numbers ... had fallen from 526,360 to 235,063 as a result of tick fever.
>
> G. Bolton, 1963:220

A few years before this the cattle ticks first came into that part of north Queensland. Before they came, Woodhouse Station had a herd of 25,000 to 27,000 head of cattle and the ticks wiped them down — on a bangtail muster — down to about 5,000 head, so Cunningham was trying to build up his herd again. He bought about 3000 head of cattle off Powlathanga Station, west of Charters Towers, and wanted them brought down to Woodhouse in two or three mobs. Dad was undecided at first, but then he said, "Well, all right. I'll do it. I'll take me horses back through Reed River up to Charters Towers and out to Powlathanga Station, and I'll bring the first mob down for yer."

After bringing the first mob down, Dad went back and brought the second mob down as well. At this stage Cunningham offered him a permanent job managing Woodhouse station. My father said: "No, but I'll take it temporary to see yer out of a hole." Well, that temporary job lasted fifty years. He started there in 1904 and retired in 1954.

In about 1906 or 1907 my father married Bridgette

Larkin and they lived together on Woodhouse Station and in Ayr until Dad retired, and then they lived in the family home in Ayr. Dad died there at eighty-six years of age. My mother was eighty-six when she died too, but there were fifteen years difference between my mother's birthday and my father's.

I was born in Charters Towers on the 17th of February, 1908, and I cut my milk teeth on Woodhouse Station. When I was five my mother sent my sister and me away to school at Bowen; I had my sixth birthday and she had her fifth birthday there at the Convent. Homesickness nearly killed me you know, and that's something I never forgave my mother for, because I reckon we were too young to have been sent away. Years later I said to her: "By Jesus, you must have hated bloody kids to get rid of us," but Mum always reckoned that she was frightened we'd grow up so wild that she'd never yard us. There might have been some truth in that too, because you know what kids are like that grow up on a station or out bush.

I had four years at the Bowen Convent and then I had another four years at the Mount Carmel College in Charters Towers. It must have had a hell of an effect on me because, even now, when I see little kids of seven and eight going away to school and they're crying — gees, that rocks me. In the Convent I was under four mad Irish nuns. By Jesus, I never forgave them either. Oh, they were hard! Everything was: "You'll cook in brimstone and fire. You'll live on and on and on, and cook in boiling brimstone and fire" — a nice thing to be telling little bush kids!

I wasn't much good at schoolwork. I wasn't helpless — I was just *hopeless!* And they'd say: "You didn't try to learn." Not try to learn, my bloody foot! I'd try and try, but I could never concentrate at school. Yet horses and cattle were just an everyday job to me. For the last twelve months of my schooling my parents got a tutor out to teach me, a fellow by the name of "Trotter" Smith, a quick-stepping bloke, which led to the nickname. He was a good teacher, but from what I can remember he was very dirty. He thought nothing of spitting on the floor and his clothes were often filthy. My mother used to do his washing for him and the sleeves of his clothes were that stiff they'd stand up.

He was a real typical old English gentleman, a remittance man apparently. A little light fella he was, and very brainy. He was a good painter and wasn't a bad man at playing the piano, and he was a great walker. Time and again he used to walk from Ayr up to Ravenswood, the best part of sixty miles. He'd camp one night on the road on the way up, spend about a week up there, and then walk back.

I had holidays from school every six months and I'd go home for a fortnight. I learned to shoot on Woodhouse when I was about ten. I had a .22 rifle for a start, then a .32 Winchester or a shotgun, and I used to shoot a lot of kangaroos and pigs and plenty of bush turkeys. If I thought there was a dingo at a certain place, I'd walk for bloody miles to get a shot at it. I wanted them before kangaroos or anything else. If I could shoot a dingo I thought I'd won a prize!

I was into shooting alligators too. I'd ride all the way over to the Burdekin River to what they called Steep Bank. The river there was about half a mile wide and you could look down from a bank about fifty or sixty feet above the water. The water only covered about a quarter of the river bed and the alligators would come out on the sand on the other side. If you want to stop an alligator dead on the bank, you shoot it between the jaw and the shoulder. That paralyses it and there it'll stop. Then you can do anything — go up to it and give it another shot if you want to, but it won't be able to move. The bullet breaks its neck, apparently.

The biggest alligator I ever got was when I was about fifteen. I got him in the Burdekin River near an outcamp of Woodhouse called Mona Park. Sid Grey, an old bachelor bloke, was there. "Have you got your rifle with you?" he said. "I'll show you an alligator down there if you want a shot."

Oh, I'd run miles to get a shot at an alligator! So Sid took me down, but the alligator wasn't there. He pointed out a great big log and said: "That's the log he lies on." It was a place where the cattle were watered — the log was only about twenty-five yards out. The alligator used to drag its whole body out of the water onto this log.

I kept going there in the hopes of getting a shot at it. One day I went down and it still wasn't there, so I walked downstream two or three miles. When I got back and looked at the log, the bugger was there. I thought, "Jesus, I've never seen an alligator that big!" I never thought they could grow that big you know. I flopped down and the first thing it did was turn a little bit — I had a big black kangaroo dog with me and the alligator was watching the bloody dog. In my ignorance I was worried my dog was going to frighten it.

The rifle I had that day was only a .25-20, a bit light for an alligator. I thought I'd shoot him from behind at an angle, so I lay down on the bank and took aim. I hit him hard right behind the shoulder and he bucked. He twisted his head around to his tail and then dived straight in. I reloaded and ran closer to see if it was going to come up again, but it disappeared.

About a fortnight later Dad and I were over at The Rocks, five miles downstream from Mona Park — the same place where Dad had met Cunningham years before. We'd ridden over to see a bloke called Bob Cox to notify him that we were going to brand at Mona Park, and while we were there he said: "By Christ, there's a big alligator dead down there in the river. Somebody shot it behind the shoulder. I measured it with me stockwhip and it's sixteen feet six inches long."

"Well," I said, "I shot an alligator down there at Mona Park."

"This one's a stranger round here, so that'll be it," Bob said.

I never saw that alligator — Dad and I didn't have the time. Bob said, "I'll tell you what I'll do, Charlie. I'll cut his head off," but some bugger beat him to it.

When I was about eight or nine I began to learn boxing on Woodhouse during my school holidays. Later on in Ayr I used to go down and train two or three times a week with some lads of eighteen or nineteen.

I always remember the first time I put the gloves on with a bloke called Stan Lennox. Stan was a Scotsman whose family migrated when he was about twelve, and they stopped in the first town they came to — that was Ayr. I was always a bit shy, coming in from the bush, and someone said, "Come on Chas, put them on." Christ Almighty, about the third bloody hit Stan knocked the wind clean out of me! That was the first time I'd ever had a punch in the solar plexus. I clinched with him and hung on as long as I could. Later on he said, "Jesus, you were hangin' on!"

"You bastard!" I said, "You bloody near knocked me with that third punch.

"Learn to keep your punches down," he said.

After a while I could hold my own with Stan and most of the others, and I've often been thankful I learned to box, because it stood to me over the years.

When I was sixteen I got a job on a ship taking horses to India. J.S. Love was the horse buyer who supplied the Indian Army with remounts.[1] My father was very friendly with Love, so I asked Dad to cite him for a job on the boat. Love said, "Yes, send him up to Townsville here and we'll give him a job goin' over." Of course, Love knew I'd been born and bred with horses — it wasn't as though I wouldn't know what to do — so I went to India with 1400 head, four deck loads of them. I just went over for the hell of it and was the youngest that ever did that trip.

We were paid £2 a week to go over and less coming back. I thought we could've been paid more — stockmen's wages in Queensland were £2-10-0 — but I suppose J.S. Love reckoned that we were getting a trip out of it. We heard that he was getting at least 300 quid for each horse that was delivered to India, but he bloody-near went broke. I suppose it's like everything else, people reckon just because you've got a big business you have a lot of spare cash. I found that out in the early years on Humbert: just because I had a station, people thought I had a lot of money, but I wasn't even on bloody wages then, I was in debt!

We left Townsville on the 7th of October, 1925, on the *SS Janus*, a British India steamship. It was a nineteen days trip over. If it had been a rough trip it would've taken a couple of days more. Those horses stood side by side all the way over and the ones that were down near the engine — Jesus, it was a big steam engine that rattled all the time — and the sweat was just running off them. Some damned steam pipes that went through our cabin gave off an everlasting bang, bang, bang! But you get used to anything if you're with it long enough, and are tired enough.

Each man was given sixty-five horses to look after and had two coolies to give a hand. We'd have to feed them four times a day and water them twice a day with buckets. I was lucky because they had hoses that year. One time you had to push the horses out of the way and clean out the dung with a shovel and bucket. This time the horses just stood while we used a big powerful hose to wash all the dung into a drain against the back wall of the hold. From there it just ran out a hole in the side of the ship and floated out to sea.

We had trouble most of the time with horses kicking at each other. They got very "gnarly" you know and oh, very sour. And bite! By Jesus, look out when you walk past them! It wasn't until about four or five days after we left port that anyone told me to look out. There was only a narrow place to walk past them and they wouldn't pull their heads back. I suppose they were a bit seasick too, but gees, if they grabbed you they hung on!

On board ship in Townsville, before we got to sea, I was doing up the grummet chain on one horse and he grabbed me by the finger and hung on. I had to punch him three or four times to make him let go, and then in my ignorance I just carried on. Ian McPherson was the boss cocky of all that — he was a nephew of J.S. Love — but he didn't know that much about bloody horses. We were the ones that knew. He passed a remark. "Look out," he said, "you want to watch them horses."

"Yes," I said, "one bit me." I showed him and he said, "My God, you can nearly see the bone there!" He took me straight up to the ship's doctor who washed the bite and gave me two needles in the backside for tetanus. Oh Christ! I couldn't sit down, I couldn't walk — that injection was worse than any damned disease I could have got. I'd have gladly let tetanus take over and died quietly!

Each horse was tied up with a grummet chain so that it couldn't bite the next one, but it could reach its feed box all right. When they kicked at each other they'd knock out the rails between them. These rails were six inches by two inches and they went into slots cut in the wood, with nothing to hold them in. To replace the rails I had to climb up behind the horses with a "hooky-lackery", a hooked stick about six or seven feet long. I'd climb up behind and hang down — if you put your foot on the rails they were likely to kick up at you — and lift the rail with the stick, and the coolies would guide the rail back into the slot. Then I had to get out of that the best way I could. A lot of the horses got used to you and you could even climb over their backs, as long as you didn't touch them, because they were all unbroken.

Eventually we got to the coast of India and I'll never forget, we were about two days in the mouth of the Hooghly River and didn't know we were in it — that's how wide it was. We got to Calcutta and then began the fun and games of unloading the horses. They were run off into chutes going down to the wharf and held in big yards there. We were about three days unloading them. Then they were sent to a depot about two miles away, in lots of 100. This depot was a big brick turnout a mile square, with walls eight feet high and two feet thick, and a big two-storey barracks inside.

After being tied up for so long those horses were very fresh, and when they were walked down the street they wanted to gallop. Four or five of us rode in front of them and a mob of coolies walked in front of us carrying a big piece of white calico three or four feet wide. This went from one side of the street to the other to act as a sort of a fence, so that the horses wouldn't rush forward and scatter. I believe a mob did get away once, and there were horses being mustered up in the streets in Calcutta for a week or ten days after.

A lot of the horses were what they call gunners. These were half- or quarter-draft horses — a heavy horse for pulling wagons or guns. A halter was put on them and we had to lead them around a square. Well you can just imagine what they were like, full of feed and not having been out of a yard for weeks. There'd be about six of us on a rope, hanging on for dear life, and these half-draft horses would be snigging us along for about forty or fifty yards before we could turn them. Odd ones got away, but somehow we managed to make them face us, and after a while we'd walk up and pat them on the forehead. Then we'd start them off and away they'd go again, and snig us around. We'd do the same again the next day. It's surprising how quiet they became. We'd give them a good brush down and we'd pull their tails — any horses with long tails or knots in their tails, we'd pull the hair out to make the end level with their hocks. It wasn't acceptable for army horses to have their tails dragging on the ground.

There were vets there, Pommy vets. I'll never forget one of them. He was a little bloke stock-full of importance. He had a little Charlie Chaplin moustache and a little cane in his hand, and you could see your face in his boots, the way he had them polished. And everything was: "I say. I say, Aussie. You bring that horse over here. I say, Aussie, did you notice any blemishes about him?" Looking for blemishes was his job, so he shouldn't have been asking us. He came up to have a look at the age of each horse. That was easy enough. He had a bit of a stick he'd put up near the horse's mouth and half the time I think he only made a rough guess. The horse would grab the stick you see, and he'd get a quick look at the teeth. "He's right," he'd say, "he's right, he's right." Actually, they were all classed according to age before they left Australia.

He went over to one horse and I thought, "By jove, that horse is going to strike." I could see that by the way it was throwing its head up in the air. The next minute it hit this little Pom in the leg and groin and spun him round, and he fell head-over-heels. He waited for a couple of the Aussie blokes to pick him up. "I say, Aussie, I could quite easily have got ruptured then. I could quite easily have got ruptured," he said. I think the Aussies were wishing to Christ he'd broken his bloody neck, they were that fed up with him.

One day Ian McPherson said, "Charlie, would you like to go down town?" I was too bloody frightened to go by myself — the bastards would knock you if they could. He said, "We'll give you a very reliable guide," so I said, "Yes, I wouldn't mind having a look at the zoo." Everyone likes to look at the zoo when they're young. They got the guide and he piloted me all through the city. As we walked he always stood behind me, and he'd pay for the tram fares so the buggers wouldn't take me down. We went from one bus onto another and at last he said, "We get off here, sahib," and finally we got to the zoo. One thing there I'll never forget. We came onto two kangaroos, a male and a female I think they were. They were very quiet and came up to me and I fed them. I never felt so homesick in all my life. I must have stopped with them a long time, because the guide kept on saying, "We move on, sahib, we move on." I looked at them and I thought, "Well I'll be going back to Australia, but you poor buggers, you'll end up stopping here." I suppose that when they died they were fed to the lions.

The guide asked me if I'd like to have a look at the fish market. The different breeds of fish there were amazing — and the bloody blowflies! There was a great big floor and I walked up to the side and looked down, and there were about 150 buffalo. Well, the bloody flies that were there! That put me off fish and meat from then on. I don't know what the hell I lived on, but I wouldn't touch fish or meat all the time I was in Calcutta. Another thing that put me off meat was when we went to a butcher's shop one night. One of the blokes said, "Hey. Get an eye-full of this." I looked and saw some three-quarter grown pups in a cage a couple of yards away. Poor little buggers, they wanted you to pat them. Then he said, "Get a look at that," and I looked around. There were three more, dead, hung up on a spike. Apparently they put them through hot water to take the hair off. Oh, God suffering bloody Christ! I nearly collapsed and I said, "What's that for?" and he said, "They eat them." Jesus, eat them!

Calcutta had a population round about five or six million from what I can remember. The streets were very wide, but it was the dirtiest place I ever saw. It was what you'd call putrid. There was no such thing as toilets, so they piddled and shat in the streets. You'd see kids about twelve years old in groups of about twenty or thirty, rounded up like a mob of cattle. Their mothers sort of bushed them and they had to get their living the best way they could. Of course if they struck anyone younger than themselves and they thought they could get two or three cents off them, they'd think nothing of murdering him for it. They camped in the streets or anywhere, sometimes just in a little hole in a wall, and you wouldn't just see one — you'd see dozens of them. God, it was frightful, terrible! Then beggars — beggars were everywhere. Oh, poverty's a terrible thing you know. The beggars reckoned if they could get one anna — that was one penny — they were set for a couple of days.

I've always remembered one kid. She was only about seven and she'd say, "Good evening, America," — she thought I was American, you see — "Good evening, America. Could you give me one anna, sahib, for my meals?" I gave her one anna and when we got about twenty yards away the bloke with me said, "I don't know how long she'll have that for." He turned around and said, "See what I told yer?" Her mother raced across and snatched it off her, and put it away. God suffering Christ! No, that really cruelled me.

That trip to India was a big eye-opener to me. It woke me up how other people lived. What rocked me was that there was nowhere for lots of them to sleep and they camped on the footpaths. When you walked the streets there were fellas everywhere and you walked right around them.

We were over there for about a month and then I caught the *S.S. Janus* back. We came back to Australia via Perth and Adelaide. There was a bus running from the Adelaide wharf and I wanted to have a look at the city, so I jumped on and went in. I walked up and down the streets a few times, but it's like everything else — if you don't know anyone or don't know how to get around, any place can be very lonely. But at least I could say to myself I'd been in Adelaide. Little did I think that in years to come I was going to make it my family home.

We got to Sydney and the *Janus* was there for days, loading jute and hides. Sydney was too wild and woolly for my liking. I was "only a bush lad" as the saying goes. I was there the day a policeman was shot and of course there was a hell of a big funeral for him.[2] Another day a fourteen foot shark caught somebody. They got the shark out at some beach and took some of the remains of the victim out of it. I was starting to get a bit cheeky then and moving around a bit. Somebody told me how to get a bus and to make sure I looked after my money and all of that sort of thing, and I went out to have a look at the shark. That was Sydney.

My sister had been in Sydney for twelve months with an aunty while she went to school there. I picked her up and we travelled back to Townsville together on a passenger boat called the *Wyreema*. It was only a three or four-day trip and if I remember rightly we arrived back on Christmas day. I went back to the Woodhouse stockcamp, working under a chap by the name of Victor Matthews. He was head stockman, and when he wasn't around, my uncle Dan Schultz took over.

Later on I had two years under Dad. I was planning to go to India again the next year but I was too late putting in my application, and missed out. It was the third year that I was accepted to go. The horse boat generally left Townsville about the first week in October, but then we got word to say that Dad's brother, Billy Schultz, had been killed in a fall from a horse on Victoria River Station.

2 Buying Humbert River Station

This is to apply for permission on behalf of William Henry Butler of Humbert River Station, Northern Territory, to transfer Grazing License No. 109 to Charles Frederick Schultz of Woodhouse Station, Ayr, North Queensland ... the sale carries 850 head of Cattle, 180 horses, Mules, goats, working gear, and the total consideration is £3500.

F.A. Brodie, 23 April 1919

Billy Butler was the bloke that sold Humbert to Dad. Butler was running horses there in the early days and then he got the job as overseer on Victoria River Downs, so he decided to get rid of Humbert. In 1919 he came in to Queensland and ran against my father in Townsville. He pitched Dad a big tale about Humbert, so Dad got his brother Albert to go out and have a look at the place. Dad couldn't make out why no message came from Albert. What he didn't realise was the big time difference between the mail and wire services in Queensland compared with those in the Territory. You see in those days the mailman only came through VRD every six weeks to two months. There were no motor cars running the mail out there then, so if you posted a letter at VRD it took twelve days to get into Katherine by packhorse. Then if it missed the train it took another fortnight to catch the next one, and if it missed the boat in Darwin it had to wait a month for the next boat.

Albert had wired Dad not to touch Humbert but the wire was slow getting through, and Dad thought that a letter must have gone astray. Eventually he decided to close on the deal with Billy Butler, and it was only three or four days later that Albert's message came along. My uncle, Billy Schultz, was managing Neumayer Valley Station at the time and Dad asked him if he would go out and manage the place. Billy agreed, and went on out to Humbert.

Later Albert came back to Woodhouse and took about twenty-five or thirty head of horses out to Humbert. He stopped there with Billy, working on the place during the dry seasons and spending each wet season prospecting for gold in the Tanami Desert. Sometimes Billy went with him.

One time at Tanami, Billy got on a bloody bull camel and rode way to buggery out. Yeah, and it was so stinking hot that the camel shed the bloody pads on its feet. Then it sat down and wouldn't get up. It was bad enough for the camel to lose the pads off its feet, but to make matters worse it was in season. Billy told me it's the bull camel that comes on heat — the cows are always in season — and look out for the bastard when he does come on. He'll be slow and sulky, and if he sees a female he'll race and hit her with his chest and spreadeagle her. The bull might not even ride her — he might see another one and go and knock her down instead.

Well, Billy pulled off the packs and his swag and everything, and put them under a tree. He had no more than a quartpot full of water, if that, and no compass either, but he was a bushman and could find his way. He waited until about half-past five in the afternoon when it cooled off a bit, although it was still pretty hot. Then he waved goodbye to the old camel and walked back about forty bloody miles.

He walked all night and got in the next day around about nine o'clock, and they reckon he only just made it you know. I remember him telling my father that he was all in when he got there. He couldn't have been too good a walker — forty miles isn't so far for a good walker.

Albert had a partner out there and they eventually struck gold, but Albert died before he could get the benefit of the discovery. No one knows for sure what the hell he died from. Some reckon peritonitis, but there was a rumour going around that his partner poisoned him with a glass bottle bait to get his share of the gold. Years later my father brought Albert's bones in and buried them with Billy on VRD.

To buy Humbert Dad had borrowed money from the bank. He was expecting that Vestey's meatworks in Darwin would provide an outlet for cattle and that this would enable him to repay the loan fairly easily, but in 1921 the meatworks closed down. Billy Schultz took one lot of at least 1000 head right down to Oodnadatta, but during most of the seven or eight years that he was there he never sent a bullock off the place — only little mobs of fifty or so to the butchers in Katherine.

Albert's death seemed to affect Billy deeply. Instead of keeping the branding irons in the fire and mustering the place properly, towards the end he was hitting the booze and running up bills. Each year the debt was increased by the annual interest bill on the original loan, but the bank didn't take over Humbert to get its money back because Dad's boss, Cunningham, had guaranteed the loan, and he stood to us.

Dad always reckoned that part of the reason he asked Billy to go to Humbert was that when he was on Neumayer, Billy was hitting the grog too much. Dad thought that by getting him to go to Humbert he'd get him away from the grog, but with the boozing that went on in the Territory it turned out to be the worst move he could've made. Billy began to hit the grog harder than ever and twice he was nearly killed in falls from horses.[1]

The Darwin police have received information from the manager of the Victoria River Downs Station, that William Julius Schultz of Humbert River Station, was thrown from his horse and died in Victoria River Hospital on 21st instant.

The Northern Territory Times 27 September 1927

One day Billy went over to VRD for the mail, and he had a few rums into the bargain. He had a blackboy with him — I think it was old Riley — and Billy started him on the road to Humbert while he went to see a drover by the name of Tom Liddy who had a camp in on the river. Tom noticed that the girth on Billy's horse was very loose. He passed a remark about it, but Billy said, "It'll be right Tommy. I'll just canter along and catch this boy. I'll catch him up along the road." He cantered away from the camp and the next minute Tom Liddy thought he heard the horse snort. He looked around just in time to see Billy hit the butt of a tree with his head. Tom had been shoeing a horse there, and whether the horse took fright and shied at the shoeing gear, or what, no one knows. Tommy ran over and found Billy bleeding from the ears and nose — the blood was spurting out.

There was no one else around there except old Fuller, but he managed to get up to the station and told them that Billy Schultz had met with an accident a couple of miles from the Wimmera Nursing Home.[2] Some men came on down there with a truck to pick him up. Billy was a big man, about fifteen or sixteen stone, but they managed to get him into the truck and took him to the hospital. Anyway, he lived for nine days without regaining consciousness, and died on about the 19th of September. They buried him at VRD.

When Billy died the manager of VRD, Alf Martin, sent a blackboy 120 miles to Wave Hill Station where a message was sent through to Townsville by morse code. No one at VRD knew Dad's exact address, so the message was sent care of the horse buyer, J.S. Love, and it ended up sitting in an office in Townsville for about a fortnight. Eventually they hunted up my father and that was the first we knew that Billy was killed.

When we got the message Dad said, "Well, we'll have to go out and have a look at Humbert." Although Dad owned the place, he'd never been out there before. So Dad and I and another young bloke called Ken Campbell went out in a thirty-hundredweight Chevrolet truck. Ken came along just to have a look at the country. We left Ayr on the 9th of November, 1927, a terrible time to be leaving those parts because we were running slap-bang into the wet season. Of course, in our ignorance we didn't know this. In those days that was the only way to get out there. You never heard of a plane doing that sort of a trip and there were no buses or anything. Everything was mostly done by donkey team or packhorse. That trip took us about nine weeks.

On the way we left the truck at Katherine while we went down to Darwin on the train, because Dad wanted to hunt up a will Billy had made. He didn't find the will, but we did get the first details about how Billy was killed. We stopped at the Victoria Hotel and one of the first blokes we met was Ivor Hall. I suppose someone told him, "That's a bloke called Schultz." Of course, Ivor woke up straight away: "Oh, that'll be Billy Schultz's brother!" He'd heard about Billy getting killed and he couldn't get over to us quick enough to tell us what he knew.

Ivor introduced himself and told us how he'd been on VRD for years. He and his brother Noel had been on Humbert before Dad bought it. In fact there's a place on Humbert called Halls Pocket which is named after them, and another place called Ivnors Pocket which is a combination of the names Ivor and Noel. When Billy Butler owned the place he got Ivor and Noel to go and look after the horses he had there. I believe that when Ivor and Noel were there they were very frightened of the bloody blacks because a white man had been speared on Humbert not long before.

Well some of the things Ivor told us, we wondered if they could be true. He said that when he and Noel first came into that country they were big lumps of lads, sixteen and seventeen. I think he said they were kicked out of Sydney and sent up to their uncle, old Townshend, one of the early VRD managers.[3] They were just jackaroos and old Townshend didn't want them there. He was a bit solid on them. "Them bloody two useless buggers," he used to say, "they don't bloody well belong to me." He'd very seldom give them a horse to ride — he'd always give them a bloody mule, a big-headed mule.

Ivor reckoned the Pine Creek goldfield was in full swing when he and Noel were first in the Territory, and on Saturday nights the Chinese would come in with their gold. (Old Noel was telling me this too). Ivor said, "Now this is without a word of a lie. They'd put all their gold together on a chamois on the table and kick off with a bloody jam tin. They'd fill the tin and level it off with a little flat ruler. 'Right. That's yours.' Each of them had their own chamois and they'd pour their tinful onto it. When there wasn't enough gold for another tinful each, they'd use a tobacco tin, and each of them got that. Then right at the last they had a matchbox. And after everyone got their share, if there was still a bit left they said to the bloke at the head of the table, 'Oh, you can keep that." Ivor reckoned that of a Saturday night there was so much gold in Pine Creek, he thought it was going to devalue, and he said he wasn't the only one — all the others there reckoned it just couldn't keep on keeping on.[4]

After visiting Darwin we returned to Katherine and did the last leg into VRD, arriving there at about eight o'clock in the evening. Alf Martin the manager had gone overseas with his wife, but the book- and storekeeper Jack Roden made us very welcome, and put us up for the night. Jack told us we couldn't get our truck all the way into Humbert but he said there was quite a mob of blacks there, and next morning he sent word over for one of them to bring packhorses to the junction of the Humbert and Wickham Rivers. The following day a blackboy piloted us from VRD to the junction where we found the packhorses waiting. We arrived on Humbert on the 8th of January.

To get from VRD to Humbert in the truck, we crossed over the Wickham just upstream from Victoria River

Downs homestead and followed the south side of the river. We went through what they call Butler Gap, where the road goes through today, and down to meet the packhorses at the junction of the Wickham and Humbert Rivers. No trucks had been over that way before. It was just a packhorse road you know, and there was no way we could get the truck across the Wickham River. The closest we could get to Humbert was about ten miles, and from there I went across with six loaded packhorses and an old blackboy called Mosquita. We unloaded at Humbert and the following day we came back for another load. While Mosquita and I were going across, Dad unloaded the truck and took it back to VRD where they let us have a shed to put it under. I made about five or six trips from the junction to the homestead, taking over different things we'd brought out on the truck. Saddles and tools — you name it, we had it.

Dad stopped with me for ten months and because of the debt on the place we decided to get things in order with a view to selling. The cattle were only half branded up, and pack and riding saddles wanted counter-lining. There were twenty-seven head of horses needed breaking in. These were starting to have age on them too, about five and six and seven years old.

As luck happened we'd brought three or four saddles out with us and they were in good repair. I'd broken in some horses on Woodhouse, so while Dad was battling away counterlining the old saddles, I hopped into breaking in the horses.

I broke in all twenty-seven and was very pleased with the job I made of them. I used to hobble them out and they'd still be flying around, till someone said, "Why don't you put side-lines on them?" — That's a chain from the front fetlock to the back of the hind leg and when they try to put out their front legs they can't do it — it brings them down onto their knees. You still had your halter on them of course. They'd fall and turn round and I only needed to do that about three or four times before they'd realise that they couldn't get away. Then you could just walk up to them quietly and rub them down the forehead, and put a bridle or halter on.

When you got them side-lined and hobbled outside away from the yard, out bush, you were getting them used to the idea of being caught in the open. One thing I can tell you, never throw a stick at a horse when he's outside and hasn't got a hobble on him. Once he realises he can run away that's the end of it — you won't catch him. You won't get within yards of him! For a start we'd lead them out with a quiet horse. Then we'd bring them back and a blackboy or myself would jump on and ride them around in the yard. Each horse got his first ride in the yard. Next day we led him outside again, and he'd be a lot quieter the second time, he knew what was wanted of him. We'd put hobbles on and we'd take him outside the yard. I used to catch hold of his mane and you could hold him then so he wouldn't bolt away. We'd canter out in front of the house for a quarter of a mile or so, and then I'd canter in the lead and the boy on the horse would come behind. You didn't let a horse straight out of the yard if there was no boy there — he'd go bush! He'd just gallop away you see, and not being used to a bridle, you couldn't pull him up. After the first couple of days they realised they had to go steady. You'd ride them out about five or six times, and then you could ride them twenty or thirty miles, no trouble at all.

While Dad kept on with the counterlining and other jobs, I did a bit of mustering round the place. I did pretty well. The first year I branded 800-odd, but if I'd known as much the first year as I did in the second, I'd have branded well over 1000. I didn't realise it at the time, but I was getting broken in myself. No doubt Dad realised the experience I was gaining.

We used to go out to Ivnors Pocket and muster there, cut out all the bush stuff, steers and bullocks, and then drive the rest, say, twelve miles in to the station. We wouldn't leave the cattle camp out there until about half-past four or five in the afternoon, which meant we'd get to the station around ten o'clock at night. We'd yard them, and next morning we'd draft them.

We only had the one old broken-down yard at the homestead. God strewth will I ever forget it? We'd put cattle in it in the morning and if they broke a rail or something they'd be gone, and all our hard riding was for nothing. A lot of the gates in the yard were broken or on their last legs, so we stretched bullock hides over them. That way the cattle wouldn't hit the rails and go through the gates. It was a good bluff too — they were up against a sort of a wall and they'd shy clear of it.

Well that was our first move. We were short of horses — we only had about forty broken-in horses there — and breaking in that twenty-seven really gave us a lift along. That was in about April 1928 and Dad was saddling all the blinking time you know, though a couple of times he came out mustering with me.

Eventually Alf Martin came back to Victoria River Downs from overseas and Dad went across to see him about selling Humbert. He wanted £8000 which at that particular time would've just cleared the debt on the place. When he came back I said, "How did you go?"

I was only a young fella then and for the first time he took me into his confidence. He said: "No good. Alf Martin will only offer us £7000."

"Wouldn't that be dropping Cunningham £1000?"

"Yes," he said, "that'd be dropping him for £1000, but look at all the money I've been putting in."

Dad had several blocks of land in Ayr, half-acre or quarter-acre blocks, and he'd sold them one by one and sent the money out to my Uncle Billy to keep the place going and keep the tucker-bags full. Well I could see he was very worried, so I said, "Don't take it. I tell you what — I'll guarantee you twelve months out here in the hope of another buyer turning up." He didn't answer me at first, but there wasn't really much choice. He said, "I'll think it over."

Around October Dad left Humbert for Woodhouse and

I hung on for the next twelve months, but no buyer turned up. Dad would never have left me there in the first place, but he really thought we were going to sell out to Victoria River Downs. He reckoned Alf Martin wouldn't want an outsider to come in there and take over Humbert, but VRD never increased its offer, and every year there was six hundred quid going out in interest on the place. In no time the debt was up to £12,000.

In the end Cunningham more or less took the place over. The bank was going to put it into receivership, but Cunningham saw the bank — I suppose he was in touch with the situation all the time. We were only getting £2 for a five- or six-year-old bullock for a start, but Cunningham had all the faith under the sun that sooner or later the place would be cleared.

Jesus, what I went through out there was bloody hell. All that had me worried was whether I was ever going to clear the place of debt or whether the damn bank was going to take over. I stayed out there with the intention of helping Dad clear the place and if I couldn't, well that'd be just too bad. As the years went by I did clear it.

Eventually there was a lot owing to me as regards wages, although that didn't really worry me. When I next saw Cunningham I had a yarn with him and he offered to sell the place to me if I could clear the debt. He said, "That's one way of getting the wages that are owing to yer." I thought about it for a long time because I'd had such a bloody pizzling with the debt, living hand-to-mouth. Then Cunningham decided to pay me three quid a week, which would just cover any small expenses — getting a bit of wire for yard building and fencing, and that sort of thing you see. And I met my own tucker bill with the few head of cattle I was selling off the place.

Cunningham had been through it himself and seemed to think that the price of cattle would rise. In World War I the price had risen to £8 and £10 a head in Queensland. Everyone thought that was marvellous and they reckoned they'd never ever see that price again. Well during the second war they were getting up to about £15 or £20 a head, and since then the prices have gotten out of hand. They're paying too much for them now.

3 "The Great Australian Loneliness"

At Humbert I suffered terribly from loneliness. The first three years were the worst. When you're living way out to buggery by yourself for months on end your mind goes over things you've done in the past, or stories you've heard. I suppose because Billy Schultz had been killed out there and I was living in his hut and sleeping on his bunk, I used to think a lot about him. One of the stories Billy had told me when I was a kid back in Queensland was about his mate, Billy Seaton. It's a long story, but it's one worth telling. I used to wonder if the same thing would happen to me.

Before Billy Schultz came out to Humbert River he was managing Neumeyer Valley Station up in the Gulf Country, and Billy Seaton was working there as a stockman. Seaton was one of five brothers who more or less grew up with my uncles. They were part Malay, but they didn't show any colour in them. All of them were pretty well-educated fellas and they knew their place, they never spoke out of turn. Billy Seaton was a good horseman, that fella — all those Seaton brothers were.

One day Billy Schultz and Billy Seaton went out to trap brumbies along the Alexandra River, about thirty miles from the homestead. They built a round yard a good quarter mile or half mile out from the water, and ran two big calico wings from it in such a position that they were off the main pad. Brumbies usually come in every day on the same track and they knew the horses would take fright at the calico if it was too close to their pad.

It was a bright moonlit night when the brumbies came

in. They let them get a gutful of water at the end of the hole and when they came out they pushed them off the pad and into the big long V formed by the wings. There were about eighteen or twenty head in this mob and it was led by a big brown stallion, one they'd seen time and again feeding out on a nearby plain with his mares. They could never get a shot at him before because he was very cunning.

When the stallion saw that it was more or less jammed in the wings it turned around and came at Billy Schultz, but they reckon Billy rode straight at him, yelling and cracking his whip like bloody hell, and he bluffed him. The stallion spun round again and tore straight up through his mares, nearly knocking a few down, and he took the lead, but he was still in the wing and it led him right into the yard.

My uncle was on the wing keeping them over and Billy Seaton was keeping the tailers coming. They'd gone a couple of hundred yards or so when Billy Schultz sang out to Seaton: "Keep the tailers comin'!" They were getting too strung out and if that happens, when the lead hits the yard and they're not pushed from behind they'll stop dead, snort, and turn back. Once they turn back they're gone — they'll come over the top of you. Well Billy Schultz got no reply and he yelled again to Seaton: "Keep the bloody tail coming, keep the tail comin'!" The next minute a saddled horse flashed past him and he knew then that Seaton had come down.

Seaton's horse got in amongst the brumbies and Billy got them all into the yard. When they get into a small yard like that brumbies will race around and hit the rails. This time, to make matters worse, they were taking fright at the saddle on the stockhorse. Billy jumped off his horse and quickly shut the gate and chained it. Then he cantered back, "cooeeing" all the time, and he rode straight onto Seaton.

Billy Schultz reckoned it was a beautiful moonlit night and he could pick Seaton out nearly forty or fifty yards away. He'd hit a hole as it turned out, and as Billy rode up to him Seaton said, "I — I'm hurt Billy. I can't move me legs. I think me back's broken." So my uncle tied his horse up and walked over to have a look at him. The first thing he did was to open Billy Seaton's fly to see if the semen had come away from him — if a man gets a broken back that's what happens, you see. The semen *had* come away.

He asked him, "Are you in pain Billy?"

"No, I'm not in pain, but I'd like a smoke."

My uncle didn't smoke so he got the tobacco from Seaton's pocket and made him a cigarette. Seaton said then, "Do yer think me back's broken?" Billy knew it was, but he said, "I don't know. You just keep quiet because I think you're hurt pretty bad all right, but you'll be right."

Well there they were, just the two of them, thirty miles away from the homestead. Billy Schultz's head was spinning and he was wondering what to do. Seaton said he'd like a cup of tea, but Billy was frightened that while he left him the brumbies might break through the gate and go over the top of him, so he said, "I'll have to go down to the river for water, but before I go I'll get you out through the wing. You sing out if I hurt yer." He put his arms under Seaton and snigged him along the best way he could for about forty or fifty yards, and he kept on saying: "Are you getting hurt? Are you getting hurt?" There wasn't a squeak from Seaton, and Billy knew his back was well and truly broken then.

When he got him through the calico wing he wanted to put Seaton's head up a little bit. He looked around and found a cake of hard cow dung and put that under him, but it didn't seem to be right. He had to go about a hundred yards to find another cake of dung to put under Seaton's head.

Seaton was talking away as though nothing had happened and Billy asked him, "What about this cuppa tea?"

"Yes Bill, I'd like a cuppa tea."

"You'll be on your own here."

"Oh that's all right Billy. If I die, I die and if I live, I live, but I'd like a cuppa tea."

So Billy cantered down to their camp to get a billycan and then he went to the river where the horses drank. He came back with the water and made a fire alongside Seaton. He sat him up and gave him a cup of tea, and Seaton kept on saying, "I got no feelin' from the waist down Billy, I got no feelin' from the waist down."

My uncle said he'd never want to go through that again. Seaton kept on talking to him. "Make me another cigarette Billy," he'd say. Billy would make him another cigarette you know, and then there'd be a long silence for a while, and he was looking for daylight all the bloody time. After a while Seaton said, "Ah Billy, now that mob of brumbies. We'll have to bring them round here so they go straight into the wing."

Billy said, "What'd you say?"

"We'll have to bring that mob of brumbies now. You watch the wing and I'll watch the tail."

Billy Schultz knew then that Seaton was going. He lived for about another hour and towards the end he said: "Bury me here Billy. Don't you shift me. Anyway, it's too far to take me in to the station."

Of course they had a dray at the station and Billy said, "I could take you in the dray."

"No, bury me here near the wing of the yard Billy. I want to hear the brumbies comin' in to water." Then he said, "I suppose the dingoes'll come up and have a look at me grave sometimes Billy? I'll be able to hear 'em howlin', and the curlews too. That's bushman's music you know. I've been listening to the flying foxes Billy. I wonder how much longer will I be able to listen to 'em."

He was saying these sorts of things, and Billy couldn't say too much — just, "Oh, I suppose, yeah," or, "I don't know." He said he'd never forget that night. Bloody flying foxes — they get in the bloodwood trees and they fight and squeal at each other. Billy said that ever after he could always see those flying foxes loping around all night while he waited for Seaton to die.

Daylight broke but Billy couldn't do anything — Seaton was dead. He had it worked out that he'd go back to the station and get a crowbar and a short-handled shovel. He'd change his horse and go back out and bury him. Billy got to the homestead by lunchtime that day and by the time he got a fresh horse and the tools, he left there round about two o'clock. He got back to where Seaton was about five o'clock and got stuck into digging the grave. By the time he was finished that, he reckoned it was about nine o'clock — well and truly dark — so he grabbed old Billy by the shoulders and put him in the grave. There was nothing he could use to cover him up so he just filled the grave in.

There he was at eleven o'clock in the bloody night, burying a man who'd died from a broken back. "What did you do then Bill?" I asked him.

"Well," he said, "naturally it cut me up. I rode down to the camp and rolled me swag out. I thought I'd take things easy, I wouldn't go home that night."

In the morning the first thing he had to do was to go back to the yard and get the saddled horse out of the mob. Billy knew he couldn't take the brumbies back by himself. He said several of them would race around and hit the rails, and then take fright and cross to the other side. At last the saddled horse came galloping round and stood by itself. It was sick and tired of the brumbies rushing and knocking it around. Billy worked his way around and caught it, and led it outside. He was worried that when he let the brumbies out his own horse might get away and follow them, so he took Billy Seaton's horse back to the camp, pulled the saddle off and hobbled it. Then he rode up to the gate and undid the chain.

He said, "I got on me horse because I was frightened they'd come over me. I pushed open the gate and then went like a bat out of hell, straight down the wing." As he went past old Billy's grave he thought, "Well Billy, you're

not in this chase." I always remember him saying that —
"You're not in this chase."

The horses came out steadier than he thought they would, but once they got around the wing of the yard they went flat out. My uncle packed up the two packhorses and the crowbar and shovel. I'm not too sure he didn't leave the bloody shovel in a tree there. And the night after Billy Seaton died he could still see the moon riding over through the branches of the trees, he could still hear the bloody flying foxes squealing and screeching, and even the bloody dingoes and the curlews let up a scream again. Later on when things settled down a bit, I think he took two or three men out from the station and put one or two rails around the grave.

> ... *After tea took a walk up to the yards & sat on rails till well after dark   came down & wrote up diary   done a bit to the horse book writing up brands.*
>
> Humbert River Station Diary, 23 October 1929

There were very few visitors and the mails took months to come and go. About the only mail I ever got was from my parents — not one of my other bloody relations ever wrote me a letter, or gave me a second thought. Evenings, I got into the habit of walking up to the stockyard with my dogs, and I'd sit up on the top rail and listen to the blacks. The blacks' camp was between the yard and the bark shed which served as my house. They had their little fires going everywhere and they'd be corroboreeing, and I'd say you couldn't have found happier people. Each night I'd wait until the moon went down — it might go down about nine o'clock or half-past. The next night it'd go down about forty minutes later with the result that some nights I was stopping up there till one or two o'clock in the morning, just waiting for the darn moon to go down. The loneliness was getting the best of me.

I'd come back to the house then. By that time of course, all the bloody blacks would be asleep, little fires burning at the head of their gunyas. I don't know why, but I'd be jittery and nervous, and I hated going into the empty house. I had a bunk made out of bush timber criss-crossed with greenhide strips, and a big bundle of grass thrown on for a mattress. It wasn't a bad mattress — it did for me, though I suppose when you're young you can take a lot. I had a mosquito net too. The gins used to come down and roll my swag out on the bunk and rig the mosquito net for me, and when I came back from the yard all I had to do was lie down. Well I'd dive into my bloody bed and the dogs would always come and camp alongside me. Curlews would often be singing out. I didn't mind them. There used to be two or three batches of them put on a bit of a turn of a night time. They're inquisitive buggers. They'd come up near my bed and I'd hear them talking, and I'd see their tracks in the morning in the sandy ground round my hut.

Sometimes it was dingoes I heard. I used to more or less look forward to them because they were breaking the monotony of the night. About half a mile over Peter Creek a pack of four or five would start up. They'd stop, and the next minute over in Muldoon Creek another pack would start up and put on a bit of a chorus there. Anyway, by the time I got to sleep it would be about two or three o'clock in the morning and I wouldn't shift till the gins yelled out: "Billycan been jump up, boss." Then I'd get up.

> *At present little care is given to the fattening of bullocks or the handling of fat bullocks. There are no fattening paddocks or bullock paddocks. The bullocks are left to fatten amongst the herd, although it is generally recognised in the industry that bullocks fatten better when segregated from the herd.*
>
> Payne-Fletcher Report, 1937

While he was at Humbert, Dad passed a remark about fencing. He reckoned Billy Schultz had told him that if he only had enough wire, or could get wire onto the station to run across the mouth of Riley pocket, it would make an ideal bullock paddock. To muster a big mob of bullocks without a yard or a bullock paddock, you'd be holding the mob and shifting them round as you were mustering each area you see. Of course, like everything else, we didn't have the money for it.

Dad knew I was inexperienced — I quite realise that now myself — and he said, "If we can get a bullock paddock up that'll be a help to yer. Before I go back to Queensland I'll have to borrow some money from somewhere to buy four and three-quarter miles of fencing." As it turned out he borrowed about forty quid off an uncle of mine on my mother's side, Les Larkin, and bought something like sixty coils of wire in Darwin. At the same time he got hold of some corrugated iron to put on the roof of the bark hut. That was the first roofing iron to come onto Humbert. The wire came down on the train to Katherine and they had the truck to cart it out from there, but oh Christ! It took about four days to do the trip from Katherine because of the damn creeks, and that sixty coils only gave us enough to run two wires across the pocket. We just didn't have the money at the time to add a third strand.

Just before Dad left, in September '28, we got a fella called Don Cronin to give me a hand to put that fence up. It took us six weeks, but that was my first experience of fencing — in later years I could have put the same fence up in about three weeks. At first the Humbert cattle didn't know what wire was. They were used to going backwards and forwards to the river, so as soon as you'd give them a start they'd race along and hit the fence and drag it down for about ten or twenty yards. Then as likely as not they'd jump back into the paddock again, so I was for everlasting repairing the fence. I could see why this was happening and towards the end of the year I managed to get some more wire from VRD. A bloke there got about ten or twelve coils and passed them on to me — enough to run a third wire. I thought I was made!

> *Arrived V.R.D. about 6 O'Clock left for the Humbert with old Riley arriving here about 3 in the morning. Learnt that some cattle perished out Riley Crk fence probably Gordon Crk bullocks ... Counted fifteen bullocks dead on Riley Crk fence & one Wickham Bull, all perished for water.*
>
> Humbert River Station Diary, 1–2 October 1929

That third strand certainly made the difference to keep the cattle in, but I perished about twenty-five or thirty head of cattle there one year. I had to go down to Darwin to see the bank in September and before I came back the damn water went dry in that paddock. That was what they call Buffalo Hole. As soon as that went dry the cattle could smell the water in the Humbert, and they came back onto the fence. Old Riley, the boy I had there, kept mustering them up and taking them back to water at the top of the pocket, but they kept on coming back.

Christ, there were dead cattle everywhere! I'll never forget that. That cut me up altogether seeing those twenty or thirty head of cattle, all dead. There were several cleanskin bulls among them that we'd never be able to muster — there they were, lying on their backs with their legs stuck up in the air. I blamed myself and lost a lot of sleep over that incident, but eventually I learnt I wasn't the only one to lose cattle on waterholes that "never went dry". I lost the cattle but gained experience.

> *Matt Wilson walked with a very noticeable limp, the result of a bullet wound inflicted during the famous 1893 Shearers Strike in Queensland ... Tho' his appeared to be a lonely life at the Depot, he was a cheery fellow and always ready to do his fellow man a good turn, there were frequent visits to his store by stockmen from near and far, the great attraction being the firewater in stock ...*
>
> C. Boulter, 1913

Towards the end of 1929 I got to the point where I was nearly out of tucker. I had no cash to buy any, so I decided to go down to Matt Wilson's store at Timber Creek to see if I could trade some bullocks for rations. I mustered up about ten or twelve pack horses and a few mules, and shod up our riding horses — it was very stony limestone country right through to the Depot.

You always get a late start on a trip like that and by the time we left the station the next day we only got as far as Police Creek, about eighteen miles up through Halls Pocket. We camped there that night and the next day went through to Bullita, about another twenty miles. From there we went through to the Eleven Mile. That's a big waterhole on Timber Creek, about thirty miles from Bullita and eleven miles from the Depot. The next day we got into the Depot round about eleven o'clock.

I went up to old Matt Wilson and told him I was broke and wanted credit, and asked if he would accept some cattle as payment. I hated asking, but had no other option. Old Matt was very deaf and when he spoke you could hear him all over the Depot. He looked at me and began to laugh. Then he boomed out: "No bugger ever offered to pay me in cattle before. I don't want any damned cattle! Just pay me when you can. How much do you need?"

"I've got seven or eight packhorses here," I said. "A couple of them could be a bit on the lame side so I won't bother with them."

"Oh be damned!" he said, "what's the matter with them?" I think he woke up then that I wasn't keen about taking all the rations on credit. Anyway, he said to one of his blackboys, "Go down and give one of Charlie's boys a hand to bring them pack horses up," so they brought up the packs.

All that had me worried was paying him back. I said, "Well, where's my rations? Here inside the store?"

"No! Take what's outside on the verandah." There must have been the best part of two to three months' supply of rations there and I said, "What! All that?" By this time I was wishing to God I hadn't come near the place because I hated debt and I was feeling frightfully embarrassed. "Oh well, the money I owe you, I'll pay interest on it." I can still hear him — he started to laugh again and he said: "No! No bugger ever pays me interest on money they owe me here."

Most of the packs were to be filled with flour in fifty-pound bags and sugar in seventy-two pound bags, ready to go next morning. The rations I wanted were twenty bags of flour, six bags of sugar, a couple of bags of rice and tapioca, fifty pounds of tea and fifty pounds of "nicki nicki" — Aboriginal stick tobacco.

When we finished packing the horses there was still a lot of stuff left over. Matt said, "When you get home, unload that stuff and send your packhorses back or bring them back again, and take the rest of this away." That's just what I did. We left the Depot Store around ten that morning and I was quite happy to pull the packs off again at the Eleven Mile where there was plenty of good feed for the horses, and a good camp for men. Next day we lunched at Spring Creek and made it to camp at Bullita that night.

Matt giving me credit for those rations meant a lot to me of course. It wasn't till around about June the following year that I took bullocks into Wyndham and was able to go back to the Depot and write him out a cheque to pay for it.

About a year after I picked up those rations at the Depot, old Jim Ronan and his son Tom were there sitting out the back talking. While they were there old Matt came past. He'd been boozing a bit and he said something about what sort of a Christmas it was going to be, and added: "It'll be the last one I'll have for a while," or something like that. They didn't take much notice of him because he had a glass of whiskey in his hand. Well Matt shuffled outside and the next minute they heard *boom* — he blew his brains out. Years and years later, when I got onto my feet a bit, I got a contractor to put up a tombstone and steel rails on old Matt's grave. He made a cement headstone and set into it a bronze plaque I had made up with Matt's name, the day and year he passed away, and my initials, CNS, in the corner.[1] I still look at that grave whenever I go past — I'll never forget old Matt for that good turn he did me.

I did that packhorse trip across through Bullita to the Depot and back many a time over the years. Once, coming back from Timber Creek, a packmule with my swag on it got away just after sundown. I think it was on Peartree Creek. It was showery weather and just on dark I said to my boy, "Bolter, pull up now and we'll take off the packs. With this rain we'll have to camp here for the night." We pulled up and hobbled the horses and I thought Bolter had hobbled this bloody mule with my swag on it. It was only a young mule, not long broken in. Well before we knew what had happened, it cleared.

By gees, that was a long night! I sat in the fork of a tree with a stinking saddle cloth rigged up as best I could to keep the rain off. There was light showery rain all bloody night and I was soaked through. I was never so pleased when daylight came. I thought, "That mule will go in towards Humbert, but it'll clear out and keep going." I got up in the morning and the bastard of a thing was in with my bloody horses. I could have kissed it!

I've seen that sort of thing happen time and again you know. When they discover the horses won't follow them the buggers will pull up somewhere and wait, and then come back to the others. I didn't take the swag off that mule — I loaded up the other packhorses and mules and drove them all straight in to Bullita where I think I camped for the night. The next day I pushed them along and went right through to Humbert. Yeah, a long trip — forty miles — and it was dark by the time I got there. After that I thought, "Well I'm not putting my bloody swag on any of these young packhorses or packmules again."

When packhorses came into a camp or homestead, it was wonderful how everyone would get excited when they heard the horse bells ringing. The gins would say, "Hello boss. You bin come back from Depot," or some bloody thing, and the dogs would greet me too. Some of the gins would grab my swag and lay it out on my bed — I suppose they thought it was a job they had to do — and inside of five minutes they'd have a fire going and billycan on. No one else to see of course. Only the bloody blacks — four or five gins running around with their kids.

I didn't have any other white men on Humbert for most of the first three years I was there. Don Cronin was there in 1928 helping me to put up the bullock paddock fence, and then I had a coloured bloke there for a couple of months, a fella by the name of Arthur Cahill.

Arthur was the son of Tommy Cahill who managed Wave Hill Station in the early days. One of my old stockboys said to me once, "You don't know 'im that Tommy Cahill?"

I said, "No. I just hear about him."

"By Christ," he said, "him little fella, but him can fight, that fella. Any big blackfella. All right. He come up, he fight'im, knock'im into shape that fella."

You were no good in that country in those days unless you could show the blacks who was boss. Yet those old boys, they thought the world of their bosses — bloody amazing, the mentality.

Arthur had been on Humbert River before my time and he used to say to my uncle, "Some day you're gonna get killed off a horse, Billy." And Billy would say, "No Arthur. You're gonna to get killed off a horse." Bugger me, they *both* ended up getting killed off a horse! First a horse ran my uncle into a tree and killed him, and a few years later Arthur got killed near Weaner Yard when a horse turned turtle on him. He lived for nine days before he died.[2]

I had no permanent employee until Roley Bowery came out and stopped with me in 1933. Roley came from Ravenswood. He got his first job with my father on Woodhouse Station when he was about sixteen, and we had a couple of years together there before I left for the Northern Territory. He was a great bloke, a good stockman and a good all-rounder. You could leave him at the station to carry on and you'd know everything would be all right. Roley eventually married a lass about five or six years older than himself and they adopted a little girl. He worked on Coolibah for a while and I think he worked on VRD for five or six months, because his wife was a school teacher there, but he was with my father and then myself for the best part of sixty years.

Roley had come out from Ayr with Jim Larkin, Archie Frew, and my father to go prospecting in the Tanami Desert.[3] They were after gold at the Granites where Albert Schultz and his partner had made their find, but a bloke called Chapman beat them to it. They went down and had a look at it and later I asked Dad, "What's your version of it?" He said: "I don't think it's what it's cracked up to be." Dad was supposed to be a first-rate prospector, but since then I've thought things over and I'm satisfied he didn't know the first bloody thing about it.

When they came back from the desert, Dad said, "We'll get Roley to give you a hand here. He's an experienced cattleman." Until Roley came I did all my own yards and buildings, because I couldn't afford to pay a contractor. I had to lay out a certain amount of money on wire and I'd get posts with a donkey team.

Round about 1937 I had Dave Magoffin on Humbert for about twelve months, and I had Dave Fogarty with me for about seven years. Dave Fogarty was only about fourteen or fifteen when I first got him. He eventually went away and got a block down at Alice Springs and did all right for himself. Dave Magoffin ended up as a well-known radio personality in Queensland. Then in 1941 I got married and my lonely times were gone for ever.

4 Outlaws

> *These tracks ... were those of Gordon "Maroun" Walgarra, Longanna and another Native ... who it is now alleged are the Principals in the Murder on the Humbert River.*
>
> Timber Creek Police Journal, 26 June 1910

> *I raided Wave Hill old Station, camped, arrested "Cockatoo" ... Cockatoo was reported to me by McIntyre, the Mt Sanford cook, to be one of the three natives who stuck him up with spears in 1924 and demanded flour etc off him.*
>
> Constable Frank Sheridan, 12 January 1926

When I first went to the Victoria River country there were a lot of blacks in the camps at the stations and outstations. At the Aboriginal camp at VRD there could be anything up to 250. I'm not too sure about the VRD outstations. Pigeon Hole had seventy or eighty, but there weren't so many at Mount Sanford because that camp hadn't been going that long you see. Moolooloo was a fairly big camp too. There'd be about forty or fifty there I suppose, but Montejinni only had twenty or thirty.

There's not that many now because they weren't breeding up and they died out. They're getting better looked after these days and they tell me they're breeding up again. I maintain that if the white man wasn't in the country today they'd probably all be dead, but of course, no thanks from the old blackfella for that — he wouldn't understand, I suppose.

There were about seventy or eighty just sitting in the camp at the Humbert — sometimes ninety — and by God I tell you what, they were a damn nuisance! A big mob like that, they'd spear your bloody poddy calves around the house and then they'd go out and clean up all the wallabies. Porcupines too, they'd get porcupines out of the hills. I've gone along there time and again and never seen a damn porcupine for six months, yet it was nothing for them to go up and bring home three or four. Apparently there's something about them that we didn't know. Some of the bush blacks in the ranges were outlaws wanted by the police for spearing white men years before. The police just never caught up with them. Maroun was one, Cockatoo was another, and a third was Possum. Cockatoo was one who'd throw a spear for a rifle shot. He'd defy Christ and all his disciples thrown in for good measure! One of the white men they were supposed to have killed was Brigalow Bill.

> *Dear Sir,*
> *I want your instant protection here at once the Blacks killing cattle and throwing spears at me. They are now hostile and defiant they forbid me to go out again I will expect you here in the course of a week.*
>
> W.J.J. Ward, 28 July 1908

When I first went to Humbert old Riley showed me a big pear tree. "You see'im that tree there?"

"Yeah."

"That's where they bin bury that Bill Ward after blackfella bin spear 'im.'

W.J.J. Ward, or "Brigalow Bill" as he was known, was the first white man to live on Humbert River. He squatted on the old Aboriginal block which is now part of Humbert, and he lived under a bit of a grass hut there.[1] It was only a block of 579 square miles. I heard a rumour that he bought fifty or sixty head of cattle, or something like that, and also ran a few horses. I think he went there because he reckoned he'd be out of the road of everyone else. He had the guts to go into that country to try and make a living and knock something out of it, but VRD seemed to be down on him all the time. They didn't want him on the place because they classed him as a poddy-dodger.

From what I could see of it he was just hunted from pillar to post, although the head stockman and others there at VRD used to look after him to a certain extent with rations and so on. My uncle Billy, Ivor Hall, and my stockboys told me different parts of his story, and I heard about it from Arthur Cahill too.

Anyway, the day Brigalow Bill was killed he was up at a yard near his house handling a young colt. He had a lubra called Lulu and she was up at the yard giving him a hand. She noticed that for the first time he didn't have his revolver on his belt — he'd left it down at the camp. So she told him, "I go down, boil'em up cuppa tea." Brigalow said, "All right. I'll finish handling this colt." When Lulu came down she got his revolver from under his pillow, put it in a bucket and took it down to the waterhole below the house. There was a big camp of bush blacks across the river from what I heard, and she told them she had Brigalow's gun. That's when they came up to kill him.

About half an hour later Brigalow finished with the colt and started down to the hut. As he was going down he saw three Aboriginals walking towards him — Gordon, Maroun, and I forget the other fella's name. They had spears, and spread out as they walked. He knew then that he was in for trouble. He stopped till they got closer, then made a rush for it. One spear got him in the arm, but he pulled it out and threw it straight back at the nearest blackfella, and he went that close to spearing him that the Abo is supposed to have thrown himself on the ground. Then Brigalow ran on down to his camp and dived his hand in under his pillow to get his revolver — it was gone.

The blacks poked at him there in his humpy for quite a while and then made as if to leave. After about an hour Brigalow couldn't see anyone around, so he thought he'd make a run for it. He wrapped a bridle around his shoulder and neck, and ran straight towards the yard to try to get a horse. He didn't get far. As he made a run for it, Gordon stood straight out in front of him and let fly with a shovel-nosed spear — got him fair in the stomach. Before they finished him off the women came and urinated on his face. Then they took what they wanted from his house and after that most of the blacks camped across the river took off up the Humbert and scattered to the seven winds.

Well of course, the Aboriginals would spear you one minute and sit down and cry over you the next — feel "properly sorry fella for yer". Lulu started to feel sorry — or frightened — and she decided to go through to VRD and tell them what had happened. She came in to the station and told a half-caste there. Well the half-castes, they have more brains than the full bloods, and he said, "What happened?"

"Blackfella spear'im that white man over there, that Bill Ward."

"Him dead?"

"Ah yeah, him finished all right. I think him finished."

"Well come on. We'll go over and tell the big boss."

"Oh no, you go tell'im."

Of course, when he got away a bit she took off and that was the last they saw of her for a while. She went down across the river and straight up on the other side.

Next morning the "yellafella" told Ivor Hall what this lubra had said. Ivor went and told old Townshend, and Townshend said, "Well we'd best look into this and see if it's right," so they went to look for Lulu and discovered that she'd taken off. Then Townshend said to Ivor, "Yer know where his camp is at Brigalow Point. Take two boys with yer and go down to Humbert and have a look for yerself — see if he's dead or alive."

So Ivor and a couple of stockboys went to Brigalow's camp and had a good look around in the hills there. There were no fresh tracks about — it looked like the blacks had been gone a week or more, but they found blood in the house and they saw where the blacks had killed a couple of steers. Ivor told me he mustered some horses that were out on the flat. He said, "Well it was no good letting the buggers stop there, so we took them back to VRD."

Ivor never slept a wink that night. He said, "I rigged me

net outside as a decoy and waited till it got really dark. Then I sneaked over and camped under a tree about fifty yards away—but I never slept." He said his two stockboys were just as frightened as he was. They went back to VRD then and word was sent to the police at Timber Creek that Brigalow Bill had been speared.

A month or six weeks after, the police came down to VRD and the chase was on. They commissioned a number of whites as special constables — I think there were about twelve or fifteen who went out to Humbert, including Ivor and Noel Hall. They went right up the Humbert River past Brigalow's hut, and made their main camp on the river at the junction of Police Creek. For years afterwards there was a big bo-yab tree there which had all the brands under the sun carved in it. Christ knows how many brands there were. It had a big broad arrow and old Ivor Hall's brand, and three or four others, including BTB, the Humbert brand. They reckon they were carved there when the police party was after the blacks.

On that evening [June 6th] *a native Known as Gordon, and who is really the principal in the Brigalow Ward murder, came to the working boys' camp at the station and speared a boy, known as Murphy, but not fatally. All hands turned out to chase Gordon but he swam the river and got away in the dark.*

Timber Creek Police Journal, 15 June 1910

They didn't manage to get all the culprits on that first trip. The blackfella called Gordon was still on the loose. Constable Tom Hemmings showed me a photo of him once and Ivor Hall described him to me. He was a round-faced Abo with eyebrows about two inches long and his eyes were sunk back into his head — little beady eyes you know. He was like a bloody snake — you wouldn't know he was looking at you. They reckon he had hair growing out of his ears and nose, and a big wild beard on him too — one of the hairiest blackfellas you could ever see. And he was pretty cheeky, he didn't care a hoot for anyone.

The police reckoned that sooner or later Gordon and the other outlaws would turn up at one of the stations and they could depend on "bush telegraphy" to let them know. Well later on a bagman died near VRD and the police came across to sell his belongings. If a deceased person had any property — maybe just a couple of pack saddles and a couple of horses — the police would sell the "deceased's estate" to the public.

Eventually Gordon decided to go to VRD because he wanted tobacco and he had a grudge to fix up with a stockboy down there — a boy called Murphy. He didn't know the police were at VRD, so he came into the VRD blacks' camp and sang out for Murphy. Not expecting anything, Murphy walked up and Gordon speared him, dropped him right where he stood. I'm not too sure if he didn't spear a second bloke, but I think the fella got away in the hullabaloo that was going on. Of course everyone scattered and Gordon took off.

When the police came to VRD they always camped about a mile from the homestead and next morning one of the trackers came up and told them: "Gordon bin spear 'im Murphy longa camp last night." The police went down and raided the blacks' camp, but of course Gordon was gone. They knew he'd make for Humbert so they scouted out and picked up his tracks where he'd gone out on a cattle pad. They tracked him across to Whitewater Hole and saw where he lay down and had a spell under a big coolibah tree there. By Christ those blacks then could walk! They'd belt out at a good five miles an hour. You have to have a good horse to walk five miles an hour.

They picked up his tracks again at the end of the waterhole. He just walked straight out on cattle pads, you see. They reckoned they'd get him because they knew which way he was headed. Gordon didn't have his lubras with him — he'd left them up at Light Creek. They knew he'd go up Light Creek to collect his lubras, then cut across on top of the hills and follow the watershed into Peter Creek, right into the really rough country. The police worked out how long it would take Gordon to get there and from what I heard they only took one horse each, one blanket strapped in front of them, and a bit of tucker. After the stir over at VRD they didn't think Gordon would go in to Humbert, so they went on to Brigalow's hut that night.

The next day they decided to take their time. They waited till sundown before leaving Humbert and got up on top of the hills at dusk. They rode across towards a spring, watching out for Gordon's fire. When they got about half a mile off they saw it. They went back a bit so the horses' whinnying wouldn't give them away, tied them up and camped. They knew that if Gordon got away he'd go straight down Light Creek, so towards daylight they sent one of the trackers ahead to wait. They told him if he saw Gordon he was to sing out: "Stop in the name of the King", or some bloody thing, and if Gordon didn't stop he should fire over his head. If Gordon still didn't stop they told the tracker to shoot to kill.

As daylight was breaking they crept towards Gordon's camp. His gins were still lying down asleep, but by this time Gordon was sitting by his fire. He looked around — he was on the alert all the time you know — and when the police were about 100 or 150 yards off he heard them coming. Of course, Gordon sang out, "Look out, pleeceman!" and took off. I don't know whether the two gins took off as well, but they reckoned Gordon yelled at them: "I'll go longa Boomundoo country". Boomundoo is forty to fifty miles further west in very rough country on the headwaters of the West Baines River. I've never been there, but over the years I heard a hell of a lot about it.

Gordon grabbed his spears and flew over into the creek alongside. He followed it down and the police chased after him the best way they could. They weren't too sure whether there were any other blacks there, but I think one of the trackers grabbed the two gins. Usually with anything like that they had the trackers broken in: "Right, if there's anybody in the camp, you stop there and you mind that mob now. We'll go round and start rounding up anybody else."

The creek Gordon got into has a low cliff that goes right down along each side. The police were one side, two trackers were on the other side, and they could see Gordon down on the bottom. He started to run down — he knew that if he came up either side he'd get caught. They reckon once or twice he pulled up and planted in the creek, but they put a shot over his head or yelled out at him. At the end they reckon he was very defiant, just walking along as though nothing mattered — till he got down within sixty yards of the police tracker waiting in the creek.

> *Gordon Sprang up and threw a spear at Jimmy a Victoria Downs boy. The boy fortunately Just dodged the deadly weapon by bowing down and causing the spear to Just miss him ... so close did the spear go that it left a streak of red Ochre in its course along his back.*
> Timber Creek Police Journal, 26 June 1910

The police tracker jumped up and sang out, "Pull up or I'll shoot," or some bloody thing. Gordon got a bloody fright when he saw someone was in the creek ahead of him. He hesitated, then hooked a spear into his woomera, walked another ten or fifteen feet and let fly. The tracker saw the spear coming, dodged, and got out of the bloody road. He sang out to Gordon again, but Gordon never hesitated. He just kept on walking straight at the tracker. When he was only forty or fifty yards off he hooked another spear in his woomera and let fly again. As he threw the second spear the tracker shot him. The policemen and the two trackers up on top had Gordon under observation all the time, and they saw the whole thing.

So that was the end of Gordon. He had guts. According to the police and the tracker that shot him, he was stubborn — sulky and stubborn — and going into things headlong, you see. They got a mob of logs and laid him on them and cremated him. They put plenty of wood on the fire and then rode in to the station. The next day they went back and got all the bones, heaped them up together and then threw on another big heap of logs and built up the fire. They never bothered going back again then because they reckoned that should have fixed him up.²

Jim Chrisp [sic] was a thin and rather pale-faced man of about thirty-five. Quiet when the action around him was quiet, he changed dramatically to a cursing and yelling demon of energy in the branding pens or on the cutting-out camps.

G. Broughton, 1965:69

James Henderson Crisp ... Manager of the Station, found natives had been killing his cattle, and when he attempted to follow them they speared him, and he almost immediately died of the wounds.

Report of the Administrator of the Northern Territory, 30 June 1920

The year Billy Schultz began as manager on Humbert River a bloke called Jim Crisp was speared on the East Baines River, on Bullita Station. Crisp was managing Bullita for Connor, Doherty and Durack. I heard details about Crisp's spearing from my Humbert River boys.

Those days they used to round up any bush blacks, bring them into the station and kill beef for them, to try and make them sit down. Otherwise they'd be out bush chasing cattle, spearing this one and that one. It's not exactly what they speared and killed, it's what they speared that got away and died that was the worry. For every one killed there might be five or ten that got away wounded.

Over a period of years Jim Crisp rounded bush blacks up on several occasions and brought them in. One evening just before Christmas he saw a wisp of smoke up the river, so he whistled up his boys and told them to bring down three or four horses in the morning. He said, "I think there's blacks up there. We'll round them up and bring them down here."

Next morning Crisp gave his boys a rifle each and they rode about twelve to fourteen miles up the river. There'd been a recent storm and they saw where a mob of cattle had been galloping. They followed the tracks for about a half a mile and came onto a beast that'd been speared, a young bullock about three years old. The blacks had taken the rib-bones and some liver out, and cut off a certain amount of meat. Then they'd rolled the beast back onto the side that they'd taken the beef from so that anyone riding past would think it had died from natural causes.

Knowing the bush blacks as he did, Crisp said, "Well, those niggers will be in there on the river somewhere. There's a big waterhole there — that'll be their dinner camp. We'll go on in." So they rode into the river, and hit the main waterhole, which was about half a mile long and fifteen or sixteen feet deep. At one end of it they saw three or four lubras in the water gathering lily roots. The lubras rushed out, but two of his stockboys rounded them up and spoke to them in their lingo: "Where *binjamans*?" (*Binjaman* is their word for husband.) They were myalls and couldn't talk English, but they pointed with their lips that the men were hunting down the river.[3] So Crisp said to one of his boys, "We'll go down that way. If we see them we'll round them up and take them down to Timber Creek, and make them sit down at the police station there." The police also used to kill cattle for the Aborigines in those days.

> *James Crisp, a well-known stockman of Auvergne Station ... was mustering cattle around Bulata Station, when he rode into a camp of blacks who had just killed a beast. He is supposed to have been talking to them for a while when the blacks speared him to death.*
>
> The Northern Territory Times, 17 January 1920

They rounded up these lubras and left one stockboy to look after them. Then Crisp went one way and the other fella went the other, and soon the stockboy sang out: "There now, they're comin' up longa river!" At the same time the Aboriginals saw Jim Crisp and his boy, and scattered. One crossed the river to where it was thickly timbered and hard for horsemen to follow. Crisp sang out to his stock boy: "You follow that bloke and I'll follow this other bloke." Apparently Crisp knew the one he chased — he'd speared two or three white men years before and been classed as an outlaw for some time, but the police could never catch up with him.

The blackfella left the river and shot across some limestone ridges — very rough going. He had about 200 yards start on Crisp who was galloping after him, singing out to him all the time, "Pull up, pull up!" and firing over his head. Crisp had a small old-time Winchester with him, as well as a revolver in his belt. As he was firing over the blackfella's head, a cartridge jammed in the breach. The story goes that as Crisp was galloping along he was looking at his rifle and trying to release the shell. The Aboriginal glanced back, saw his chance, and whipped behind a big tree. In about ten or twenty seconds Crisp was past him. He'd only gone a few yards when the Aboriginal let fly with his shovel-nosed spear and hit him in the right kidney. From what I was told, Crisp dropped the right rein and was half-out of the saddle on the right-hand side. He was pulling on the left rein, which meant that he was pulling the horse away from himself all the time. He went on like that for about seventy yards in a half-circle. At last he fell off, and as luck happened the horse kept going and galloped straight down the river.

While this was going on Crisp's stockboy had another blackfella bailed up, but when he saw the riderless horse he didn't bother any further with him. He galloped across to stop the horse and of course he was wondering then what happened to Crisp — whether he'd been speared. He tied the horse up and went back on its tracks, and he came straight onto Crisp. He was lying on the ground with the spear nine inches into his back.

According to my boys, Crisp said, "Take my hat and get me a drink of water." So the boy galloped into the river and brought back a hatful of water, but when he got back Crisp was dead. You can imagine the panic he'd be in then, wondering what to do. He went back down to where the other stockboy was with the gins. They let them go and decided to carry Crisp to the river — he was speared about half a mile out. They covered him up with bushes then and rode back to Bullita.

Crisp had been the only white man on the place. At Bullita one of the stockboys got a fresh horse and rode in to the police station at the Depot. There was a bit of a bridle-track from Bullita to the Depot those days, but he wouldn't have followed that. He'd just cut across country. That stockboy did a mighty ride that day. He rode out about fifteen miles that morning, then fourteen miles back to Bullita, and another thirty-odd miles in to the police station. That's more than sixty miles.

He arrived at the Depot around two in the morning. In those days the police always had dogs. As the boy rode up to the house the dogs came out barking. He sang out: "Jim Crisp bin get speared by blackfella."

"When'd it happen?"

"Oh, dinner time today."

"How far from Bullita?"

"Up the river. Sun bin shovel-handle high." Shovel-handle high is round two o'clock; axe-handle high is about four o'clock.

The policeman asked, "Is he dead?"

"Oh, him dead all right."

"Where is he now?"

"We bin carry'im and we bin put'im longa river, cover'im up with bushes."

> Constable Turner will probably have black trackers with him, but will have very great difficulties in food and transport over rough country in running down the murderers. He is an expert bushman, however, and understands the native language, which should pull him through with safety.
>
> The Northern Territory Times, 17 January 1920

The police always had about fifty or sixty horses in a paddock about four to five miles from the station. Well they had to muster them in and then start shoeing up. By the time everything was ready, they didn't leave Timber Creek until the second day after the spearing. The police always took about three trackers with them in case of arrest — if they arrested the culprits they'd have to arrest a number of witnesses too.

They went to the Eleven Mile, a big waterhole on Timber Creek, camped there that night and the following day went through to Bullita. Next day they hit Jim Crisp's body round about lunch time. Of course Crisp, he'd be pretty high by then — that'd be about the fourth day he'd been dead, you see — so they just dug a hole alongside and rolled him straight into it.

I think it was later on they caught the murderers over near Auvergne. Those bush blacks were on the move all the time.

> *An attempt was made by police to apprehend the offenders, who resisted, and one was shot dead, the others escaping into the hills. The search is being continued, but, as the country is rough and horses cannot travel amongst the hills, the search must be on foot. I fear non-success ...*
>
> Report of the Administrator of the Northern Territory, 30 June 1920

> *I was speared by Blacks last night at Bullita Stn it was one of the Bullita boys that did it a gin went away for a day & during the night the Boys came & told me she was in the creek but would not come up so I went to tell her to go up to the camp when up they jumped with spears and speared me in the back, but luckily I ran for it I had no firearms with me.*
>
> Harry Condon, 31 March 1910

I used to know an old bloke called Harry Condon who'd been the manager of Bullita in the early days. Harry was a little fella, only jockey weight he was. He used to put in a lot of his time down about Borroloola and in his old age he lived on Muckety Station where Fred and May Ulyatt looked after him. Harry got hit with a spear at Bullita too. He told me about the spearing himself.

Harry was at Bullita in 1910, ten years before Jim Crisp was speared. He said he was looking after the place and one day a gin came up from the bottom of the river and said, "Boss, one lubra down there, him sick, down the end longa waterhole."[4] Harry reckoned you were always on the alert those days. "What's the matter with him?" he asked.

"Oh, I don't know. Him sick."

"Well go down and bring her up here."

"Him too sick, can't come."

Down below Bullita homestead at the end of the waterhole, there's a lot of rock. Harry asked, "Where is she?"

"Oh, him down longa rock longa waterhole."

"All right. I'll come down and have a look."

So Harry buckled on his revolver, whistled up his dog, and walked on down. He told me he only got down past the house about 200 to 300 yards — there's a bit of a blind gully with boab trees around it that comes in there. As he got more or less opposite one particularly big boab, an Aboriginal with a spear jumped out in front of him. Harry reckoned he got such a hell of a shock he just pulled up in his tracks. The blackfella was fumbling around with his spear and just as he was about to throw it, Harry's bloody dog rushed at him and threw him off-balance.

Harry still got hit — he showed me the scar where he got it right through the bloody arm. I asked him what he did and he said, "I pulled it out. I didn't know how many blacks were there and I didn't wait to see." Apparently there was only one, but Harry said, "I hit me straps for the house as hard as I could go." And while he ran the cattle dog took after this bloody blackfella and heeled him up while he ran around in circles. Well Harry said it gave him a hell of a fright, but he got through it all right and it was the best thing that could've ever happened, because it made him a lot more careful you see.

About '97 a man working on a Cattle boat Called The John Hockings got left at Wyndham his name was Jack Newton ... Newton told

> *Con Keeler that he had been in america Keeler Nicknamed him "Colorado" ...*
>
> W. Linklater, Manuscript, n.d.

> *Started out mustering last Sunday afternoon, in company with boss going up to Bullata, and two drunken men ditto. One, "Colorado" Jack, persisted in galloping in all directions, trying to bump the other riders, cutting out horses out of the mob, and other silly tricks. Strange that he did not fall off. He jerked about the horse from tail to ears, lost both reins ... he did not fall off.*
>
> Diary of J.R.B. Love, 1912–14

It wasn't always blackfellas knocking down whitefellas either. I remember Frank Spencer talking about a bloke called "Colorada Jack". Colorada might have been one of the blokes that murdered the bloody blackfella there at Bullita. It was before I came to Humbert of course, but they were still talking about it when I first got there.

From what I heard it was a grog turnout. They were full, and apparently the blackfella was pinching their grog or some bloody thing, so they knocked him on the head and killed him. Then they threw the bloody axe in the river. They didn't mean to kill him, but on the grog things can get out of hand.

Somehow word of the murder got to Tom Hemmings when he was on patrol at Auvergne, and Dan Carrol happened to be there when the message came through. Dan was the bloke who put up Carrol's Yard on Kaiser Creek, over on VRD. He was a big raw Irishman and a mad wild bugger too, but he turned out a pretty good man in the bush, seeing he was a "townie" to begin with.

Dan heard Hemmings say he was going to head back to Timber Creek to change his horses and then go to Bullita to arrest Colorada Jack and his mate for murder. So as soon as Hemmings left, old Dan rode straight up the East Baines River. It's fifty miles from Auvergne to Bullita and by gees, it'd be a rough ride you know. I think he did it in one day too. When he got there old Dan said to Colorada Jack and his mate: "Clean up 'cause 'emmings will be 'ere ter-morra." Then instead of getting to hell out of it, he stopped there and old Hemmings turned up the next day.

Tom Hemmings was an officious sort of a bloke — "I am a policeman and that's it, and I'm here to do me job." He first came up in that country to work in the post office at Katherine; I think he was there for a couple of years before he decided to go into the police force to make more money. I remember him telling me that he hoped to retire with enough money to buy a small farm in South Australia.[5]

The police could make good money from the Aboriginals. Any Aboriginal prisoners and witnesses they brought in they had to feed. They were allowed so much per person, I think about three shillings a head which was a lot of money in those days, so they'd get as many witnesses as they could. All they fed them on was beef of course, and they'd get their beef for free. Being the police they didn't ask the station. They just went out and blew a beast over. Then they charged the government. Police could get away with a hell of a lot those days.

Anyway, Hemmings turned up there at Bullita and the first bloke he clapped eyes on was old Dan Carrol, and he knew right away: "That mongrel has told them I was coming!" Well! Didn't Hemmings blow the bloody shitter out of old Dan left and right, but there wasn't much he could do about it. They reckon Dan just sat back on his swag and said, "Oh, I wouldn't bother about that Tom if I were you. I think yer worryin' too much about this, Tom." Yeah!

Hemmings had a tracker there and they questioned the gins, and one said, "Oh, they pick'em up that axe and one a them white man bin chuck'im right longa middle longa water." They walked down to the river to an opening in the pandanus and the gin pointed out where the axe fell. So bloody Hemmings told the tracker, "You go out and dive down and see how deep it is." The gin told him how far out to go. She said, "Yeah, little bit more far, keep goin', keep goin'. Might be there now, I don't know." Down the bastard went and you know, they reckon the tracker pulled the axe up first dive. First bloody dive they reckoned! Talk about luck! Well they took them to Darwin for trial, but they got out of it through lack of evidence.

> C.N. Schultz reported finding the remains of fourteen head of cattle in one pocket on his station, that are believed to have been killed by aboriginals their heads being battered in, in most cases.
>
> Timber Creek Police Letter Book, 31 March 1929

We didn't get to Humbert till four months after Billy died, and of course the bush blacks heard about Billy getting killed, so they just came in and took over the place. There were about twenty or twenty-five in the mob. Actually, a lot of them were ex-stockboys. Tommy Cahill and all those old hands had them on horses and certain ones that were all right they kept, others they sent bush. We were never short of labour in those days because the blacks were just raring to get in the stockcamp, glad to get away from bush tucker and that sort of thing. They were happy people back then.

The bush blacks played havoc with the cattle. Everyone reckoned that for each one the blacks got, about five or six were wounded with a spear and got away to die you see. They made the cattle really nervy. You only had to let out a yell — "Hoy!" — and holy ghost! Wouldn't they scatter? They'd take fright you know and you'd see them run with their heads in the air. If you rounded them up and drove them along on horseback, after a while they'd steady down.

Well the beggars were up in Riley Pocket when we first got to Humbert. We heard they were there and we sent word to them to shift — you never let them stop where your bullocks were of course. Not long after Dad left I rode up the Riley Creek bullock paddock. I had a big lump of a twelve- or thirteen-year-old kid with me — we used to call him "Humbert Charlie". We rode right up to the head of Riley Creek and just as I was about to turn back I saw the remains of a fire. I rode over to the ashes and saw bullock bones there, and I thought, "Hullo, the blacks have been spearing cattle here." Then about fifteen or twenty yards away I found another fire, and another. I found four, five, six, and I thought, "Oh Jesus!" I woke up straight away when I saw the fires right across the pocket. At this point Riley Pocket narrowed in to only about fifty yards across. Further up it widened again and continued up into the ranges for about a mile. The blacks had camped in a line across the pocket to stop the cattle from coming through during the night, to keep them back where they could easily spear them. Then whenever they wanted meat they'd just go up and, bang — knock one over.

There was a creek up on top of those ranges and about a forty-foot drop where the water came over, straight into Riley Creek. Well I followed the creek right up till it got too rough for my horse, and I thought, "I'll go up there and have a look where the water comes over. It wasn't running at that particular time, but there was still plenty of water down in the creek. I dismounted and tied my horse up, and only went about ten or fifteen yards before I realised that my leggings and spurs and elastic side boots would be a hindrance climbing up the rocks. I kicked them off and threw them back near my horse.

I followed the creek bed half way up the side of the range to a cliff about sixty or eighty feet high, and just as I was about to turn back I noticed something in amongst the bushes and big boulders. I looked and I thought, "There's a damn dead beast there!" I walked over and that's what it was all right. It was a dead "speary-horned" bullock that would've been about eight or nine years old. He'd been very fat — a fat bullock dying like that, the oil spreads out six or eight feet right around him. He'd been speared down on the flat somewhere and must've been hit pretty hard, but he'd got away from them and run up there and got in amongst the bushes, crook and sulky, and he lay down and died where I found him. I was young and inexperienced those days, but I've often thought since that if I'd had a look I probably would've found the spearhead still in him.

Anyway, I went back to my horse and on the way back down the bullock paddock I saw another dead beast. This was only a young bullock about three years old. Humbert Charlie was with me and I said, "What d'you think — speared?" He got down and had a look and said, "Yeah, spear bin hit'im longa neck." They didn't get this bloke either. He'd probably been hit somewhere higher up the pocket and run away, and he just lay down and bled to death, apparently. There were nineteen bullocks killed there that time.

> *Went right to the head of the "Light" saw several niggers tracks otherwise saw nothing of interest.*
>
> Humbert River Station Diary, 21 June 1929
>
> *Left for Bamboo springs ... Saw a dead bullock junction of Police Crk speared in the guts by blacks & just died.*
>
> Humbert River Station Diary, 2 August 1929

I never saw those blacks the police didn't catch, the ones that had speared Brigalow Bill and Crisp. They stayed for good out in the ranges, but when I think about how I used to ride around on my own in that country, I shake! Along Light Creek and around the junction of the Light and the Wickham, they'd come down the Wickham River there sometimes and on a couple of occasions I came on their tracks.

Old Riley always used to tell me, "You look out, Charlie. That Maroun, him properly cheeky bugger that one." Me being young and silly, I'd say, "Oh? Yes, all right. He's cheeky, but I've got a rifle and revolver here." If those outlaw blacks happened to see you or hear you coming, they'd just get behind a tree and you'd ride right past them. It's the bloody spear you don't see coming that's the trouble. The one you see coming is no faster than a cricket ball and nine times out of ten you could get out of the road, you see.

I came onto their tracks at the Light once and saw where they'd speared a three or four-year-old cow. It was only weaners they hunted. I had a feeling at the time that they had a dog with them and the bloody dog would pull down the cows for them. They'd only take a shoulder off a weaner and maybe the liver too — they liked the liver. I often wondered what happened to those blokes. I suppose they died of old age out in the bush.

5 The Big Run

> *This is a fine big property capable of running 90,000 cattle. It is composed of a big area of first class well grassed country in the south, through good and useful country to hard light carrying country in the north-west corner. It is infested with wild donkeys and unmanageable scrub cattle and some buffaloes but not many, no brumbies. Nearly all the country carries heavy surface stone and because of it I believe is one of the hardest places on horses I have seen.*
>
> Report on Victoria River Downs Station, 1959

From the time I first went to Humbert I had a lot to do with Victoria River Downs, or "The Big Run" as it was known. I always reckoned I was the underdog in that country. Well I was, too, especially with VRD being the "big place" you know. Whoever was the manager of Victoria River Downs, what he said "went".[1] Some of those bastards were like policemen, but when it's all said and done they were just employees. Christ, the way that bloody station got murdered! Honestly, I shamed the bastards, and it was hand to mouth for me. The worst thing that could happen for a young fella starting out was to go to VRD where they were slap happy. If they lost ten or fifteen or even twenty head yarding up, well, that was nothing — VRD had mobs of them. But on a little private place like Humbert, every calf's a calf to you, if you see what I mean.

A lot of the VRD managers were hopeless. I'll give you a couple of examples: On Montejinni, Snowy Shaw would muster 1000 or 1200 bullocks to go to Queensland. The bullocks he mustered might be on the other side of Montejinni in towards Murranji, somewhere around there you know, yet one manager would make Snowy bring them all the way back to Dashwood drafting yard.

In due course they were bang-tailed and innoculated for pleurisy or "pleuro", and then handed to the drover. Now, from Dashwood yard to the other side of Montejinni would be the best part of eighty miles — eighty miles up and eighty miles back again. They'd be walking for five or six days before they got to Dashwood. That meant those bullocks had an extra ten or twelve days on the road, losing condition.

That same manager, if a yard got burnt down he'd never think of putting it up again, and he went for open broncoing. He said, "Oh yes, catch the buggers open bronco." Jesus, that bloody bronco work used to give me the shits any time, and when it came to open bronco, forget it. They'd have about fifteen or twenty blokes rounding up the cattle, trying to hold them in a mob. They'd lasso a calf, pull it up to a tree or a fork in a tree and brand it. If it was a male they'd castrate it and then they'd let it rip.

A calf just branded or castrated would still be hot and often it'd race straight back to the mob, and it would just as likely keep going out the other side. That would disrupt the other cattle and make it hard for the blokes on horses to hold them on the camp. Cattle would be continually trying to break away and rush, and the horses did nothing but gallop, gallop, gallop, all the time. Talk about a horse-killing game!

Those were things that this particular manager used to do you know. It used to tickle me and of course the other blokes would talk about it too. Magnussen was just as big a bastard. If ever there was a bloody snob came on to the bloody place Magnussen was one, and a sheep man into the bargain — put onto a cattle station! He was a jackaroo, a raw jackaroo from around Tibooburra, and he came up and spent eight years there as manager![2]

That's the way VRD was getting murdered all the time. He was there eight years and by the Jesus, you should have seen some of the things he used to do! I was married then and my wife said to me, "Charlie, he's getting information from you." I said, 'Oh, nonsense!" but it suddenly dawned on me that the bastard was always asking my opinion — he used to keep on pumping me about things.

Scott McColl was all right but he was only there for about twelve months. Old George Lewis was all right too, but I kept out of his road, I never bothed him. Ian Michael was good — a very fair man, Ian.

> *A capable first class cattle man with years of experience. By sheer ability he rose from spare boy to stockman, drover, head stockman and manager ... Almost illiterate he writes stiltedly and reads about as [well as] a primary school child. He knows his job inside out and [is] up to every occasion, coming from a poor family of very Irish stock. He has proved every lesson he learned and pigeonholed it in his mind. This has made him inflexible in addition to his stubborn [sic] nature. He dislikes change and anything however good finds no favour with him unless he has proved it years ago.*
>
> Description of Jack Quirk: Report on Victoria River Downs Station, 1959

Jack Quirk was the mongrel Magnussen would like to have been, but Magnussen was a bit on the frightened side — he didn't know how to take me. Quirk was a domineering sort of bloke. I had no time for him, so I just kept to my bloody self.

Old Quirk had Jack Farquharson building yards on VRD one time. Jack was a big yellafella, a good yard builder, but he was a little bit soft — you could talk him into anything. At one stage Quirk cut right down with the yard building on VRD, so Jack came over to me one day and cited me for a job. "Any work here?"

"Yes," I said. "I want some yards put up, but you'll have to put your time in with Quirk."

"I've put it in," Jack said.

He came to Humbert with his gin and I put him on to building wire yards — three, I think it was. He used to put one up in nine days and I paid him a hundred quid a yard. While Jack was with me Quirk wrote to him, but old Jack couldn't read too good, so he asked me what it said. Oh gees! I shook my bloody head and I said, "Don't you tell Quirk that you showed me this letter." In it Quirk said: "I'm afraid you won't be getting as well treated over there on Humbert as I'd treat you Jack, and last but not least, your kid — I'll see that you get your kid back. I'll send away for her the day you get back here."

Jack Farquharson's little girl had been taken away by the government and was going to school in Alice Springs. Gees, Quirk wrote a rave and old Jack asked me to read the bloody thing! Quirk should have woken up that old Farquharson couldn't read, but that was typical of Jack Quirk. I never forgot that.

There was a lot of competition between the stations. I suppose there wasn't much to skite or talk about and the only fun and games you had was what you made yourself. I got caught out with bragging one time. We'd finished mustering and I was going to take some cattle in to Wyndham. There was one plant of horses in the mustering camp, but we needed to shoe up another plant to go to Wyndham. We were running a bit short of time, so I slipped on ahead into the Humbert and mustered the horse paddock there. All my boys were still out in the mustering camp, so I got a couple of lubras to give me a hand and I began to shoe the horses.

I tied the horses up and got the lubras to bring me about four at a time. I'd shoe them and then hand them back to the lubras. They'd hang a pair of hobbles on a strap around each horse's neck, then let it go in the horse paddock. It was routine work and I got the last shoe on the fifteenth horse around about half-past five. Then I had to ride out another twelve or fourteen miles back to the main mob of cattle.

Well I used to pat myself on the chest when I was telling the VRD mob about the fifteen horses that I shod in one lot. Normally, if you can shoe five or six horses a day you're doing pretty good work. I was doing a lot of skiting until

I heard there was a coloured bloke at Wave Hill who shod *seventeen* in one day, and every time they saw me the Wave Hill mob didn't forget to remind me about it either! They were always taking a rise out of me: "Have you beat that number yet Schultz? You reckon you shod fifteen. What about the seventeen, have you beat that yet?"

Back in the '30s, we'd have a bit of a sports meeting at VRD on Christmas day. There'd be about twenty of us in it, all told. There was a high jump and broad jump, the hop-step-and-jump, a ladies' race, and of course they had to have a hundred yards sprint — the "Stawell Gift". Usually the VRD mob liked everything all VRD. They didn't mind VRD blokes belting each other in a fight or a race, but when an outsider beat them they didn't like that at all. But this time, when the sprint came up they said to me: "Come on Schultzy, you're in this." All the events had prizes and the prize for the hundred yards sprint was a beautiful pillow-case made by the Sisters at the hospital.

Well we got ready for this bloody sprint. One of the blokes in the race was Jack Knox. I liked Knoxy. He was a good man, a thickset sort of a bloke — you wouldn't think he was a runner. There were ten of us lined up for the start and I remembered something I'd learnt at school sports meetings. At school there was always a starter and nine times out of ten everyone would be looking straight ahead, waiting for the starter's signal, but one day I woke up: "*Look at the starter*. Never mind about waiting to hear him say 'Go!'" As soon as I'd see him start to swing down his flag I'd take off, and I beat the others out of the holes every time. I was watching the starter now, and as soon as he started to swing his flag down I never waited for him to say *"Go!"* — I was *gone*, and I got at least a yard start on them. Away we went and in no time Knoxy drew up alongside me and there he hung. Of course a lot of them were mugs who just went in for the fun of it. At the last of it we were all strung out for the best part of twenty yards, and Knoxy and I came in together. I had a feeling that I'd won, but a VRD bloke was the judge and he gave it a draw.

They wanted Knoxy and me to run it again, just the two of us, but I said to him: "Well look, I beat you out of the hole, but I feel a bit guilty about it," and I told him my trick and gave him the pillow-case. "Gees," he said, "I knew you were fast out of the hole!" The two nurses who'd made the prize weren't happy about it. They said: "No Charlie, that was for the winner. You should've turned round and run it off with him."

They had several races for the boys and several for the lubras, and gave them prizes — shirts or dresses or something. We'd line the gins up and by gees it was funny. You'd say, "Go on now, put on your trousers." They'd put on their knickers and by Jesus some of them could run! One race we had was the three-legged race — they'd get the lubras onto that too — and then they had the bag race. By Christ, you didn't care who the hell won — it was just seeing these people tumble head over turkey that was so funny.

And for the kids there were lollies by the truck load! Lollies and soft drink. One thing the kids got going at the sports meetings was the treacle-and-bun race. It was mostly for blackfella kids. They got a lot of buns and tied them so that they hung down from a rope. Then they smothered them with treacle. They'd stand there with their hands tied behind their backs and someone would sing out, "Right, go!" and whoever ate their bun first got the prize.

And then there'd be a big booze-up. Jesus Christ — they drove me mad! My mate Roley would have the odd drink, but they used to try to get him drunk. I'd say to him, "Look out now, you don't want to start gettin' bloody-well full with that mob there," and he'd say, "You leave it to me. I'll just have one or two and as soon as I feel as though I'm starting to get merry, I'll leave 'em to it." It was all hardtack you know — whiskey and OP rum. There were usually several bottles of brandy and gin for the women, but if all the other grog cut out the bloody blokes usually weren't too proud in coming forward and getting the brandy off the poor women.

Everyone seemed to go for the grog. I suppose it was a part of "The Great Australian Loneliness" and this was one way they met it. Of course white women were few and far between then — for every one you'd see there'd be about fifty men there.

A lot of the grog on the stations was brought out by the mailman. There'd be around about forty packhorses come out and I suppose twenty of them would be mail and the other twenty would be grog. Not beer — it was all hard stuff, because you couldn't carry enough beer. When it turned up there it was one glorious booze-up you know, and I could see what was happening, so I just kept to myself. Once they realised that I didn't bother about boozing they left me alone.

We used to have Christmas race meetings too.[3] The first I went to was back in 1933 — I'll always remember that year. They were held on the long red flat near the present-day homestead yard and they were just straight sprints. All the outstations came together for them and had a great time. Jack Roden used to run a book and by Jesus, three and four to one was nothing for him to give out. He was the bookkeeper and storekeeper on VRD for about twenty-five or twenty-six years. He was a likeable sort of a cove, Jack Roden. A bugger to booze, but it made no difference — on Monday morning at the tick of seven o'clock that store door opened. We all thought he was going to get the management of VRD. He'd have made a good station manager that fella — bloody oath he'd have made a good one! He was always on the ball and could write a good letter, and he knew when VRD was getting a good day's work from the head stockmen. But eventually a new manager came there who realised Jack knew too much — he was a threat to him — so he asked him to retire.

Everyone came along to the meetings. Roley Bowery would ride some of the horses for me and Bob Nelson from Pigeon Hole would have horses there. Snowy Shaw would come over from Montejinni, Jack Knox from Moolooloo, George Bates from Mount Sanford and Frank Spencer from Gordon Creek. All the horses were put into the yard at VRD and auctioned off and whoever bought one only "owned" it for the day. Part of the money they paid for the horses went to the Australian Inland Mission Hospital at VRD and part went for prize money.

At one of these meetings, when the horses were being sold through the yard, Frank Spencer came around to us and said, "What have you got there?"

It was a mare called Biddy. I said, "Well, that's our fastest horse."

"What's she like?"

"Oh, I don't know about her winning, but she's very lively on her feet."

Roley and I were broke — we didn't have any money to be paying out for horses. When Biddy came through the yard Spencer ran us up to about four quid. That was all the money we had, so he got her and I said to him, "Oh gee, I hope you start her in the bracelet."

Well Spencer started her all right, but he was no sport you know. If he could win any races he'd win them. Others he could have let win a race there, but not on your sweet life! Frank Spencer and Bob Nelson were great mates, thick as thieves, and Frank wanted one of Bob's horses to win, so he tried to rig the race. He put a half-caste "Chinee" called Mahnikee on this horse of mine you see, and he said to Mahnikee: "Now if you're gonna win, you pull up. Don't you win now."

The race kicked off and about sixty yards down on the straight Mahnikee was in the lead, but he was looking at the other riders behind and beginning to slow up. I looked and I couldn't believe my eyes! I yelled out, "Get a move on you bastard, get a bloody move on will yer?" Mahnikee saw me — lucky that he happened to look my way — and he kicked his horse and won by about half a length. Jesus, wasn't Spencer crooked over that! I said to him, "Serve you bloody well right you bloody bastard" — I abused him for everything you know, but it made no difference to Spencer. He didn't care a hoot. I said, "To think Roley and I only had the one horse and we only just came over for the fun of it, yet you turn around and try to win in that underhanded way."

We came down on a different occasion, the time we won the bracelet on a horse called Dave. Jesus, listen! We cleaned things up that time too, because the first horse we started was our worst, yet he won it by miles! Roley said, "What the hell's going on here?" I was amazed too, and said, "It's not as though he just won it. He won it by about four or five lengths!"

A blackfella's race was next. They used to like to race too, and there'd be about three races for them. In the next race one of our horses won again! Roley looked at me and said, "We won the VRD Cup that time!" We won it, but good old Frank Spencer lodged a protest — that was typical of him. I don't know on what grounds Spencer lodged the protest, but as soon as he did Roley got over to me. "Now you get to bloody hell out of this," he said.

I didn't know what was going on so I said, "What's up?"

"They've lodged a protest. Just get to hell out of it and I'll fix it."

He reckoned I was too bloody quick-tempered and I'd end up making a bloody ass of myself, so I went and sat under a tree. Bob Nelson didn't have anything to say, he was quiet about it — it was Spencer doing all the bellyaching. Alf Martin was the judge. He came over and walked around the horse a couple of times, and saw the BTB brand on him and said, "It's a Humbert brand and that's all there is to it."

We won five races and I think VRD might have won two that time. In previous years I'd walked my horses through Bullita to the Depot races, just for the hell of it — it took about four days to get down there — but I never won a race. My ambition was to win just one, then all of a sudden the "Wheel of Luck" turned for me and there you are. Good times? Oh Jesus, I'll say they were!

There, lording it in the biggest of brush shelters his natives have put up the day before, Charlie Schultz dispenses hospitality like a feudal baron of old. He is host to police, a trio of pretty nurses, and anyone else who happens to come along ... A dance at night wound up the first day's events, and some of the local boys came up with entertainment which was enjoyed by all.

K. Willey, 1964:29

In the 1950s and '60s, races on VRD were held about two miles up the river, right on the Wave Hill Crossing. Magnussen started off holding the race meetings there. Later on George Lewis became manager — he was a racing man who believed in doing things properly. We had bough sheds there roofed over with spinifex grass and by gee, when it's put on properly it's there to stay. We used to have a race meeting there once a year. It was all bush horses and they were grass fed — nobody fed them grain or anything. When you start having to feed them you lose too much time for a start and you're supposed to be working, not feeding horses. I just got one or two of the blackboys to trot them around in the morning and then let them go down the paddock of a daytime.

Pigeon Hole always had a strong team there with five or six horses, and Mount Sanford usually had four. The Centre Camp would have at least three or four, and Montejinni usually brought over four too. I used to feel sorry for Montejinni because their horses weren't the best, but still, I went there for about three or four years running and never won a race. My one ambition was to run a place. Then there was Moolooloo. They always had four or five horses, and Gordon Creek the same. I generally brought over five, and Wave Hill always brought over five or six. On several occasions there were horses from around the Depot. Some even came from Wyndham.

The races were always run in about the middle of the dry season and we always seemed to strike the very cold weather, with the result that George Lewis used to pull in big logs and put them around the meeting area, and light fires. He'd start the fires round about five o'clock in the afternoon and by night time they got a go-on, and the place was warm.

A hell of a lot of people used to camp up at the Big House at VRD. Down where we were, right on the banks on the bottom of the river, there were four or five different camps and I suppose in each camp there'd be about six people. There were always two police turned up there and a couple of planes always flew out from Katherine and Darwin.

There was a good concrete dancing floor too, and there was a dance every night. Everyone would get out on the ground and kick their shoes off. The women danced around on their bare feet. Oh, the dust would fly and they were never so happy in all their lives you know. The last night when the prizes were given out there was always a big feed-up and a bit of a booze-up, though not too bad on the whole. The prizes were trophies and cash, and by gees I tell you what, the prizes were very good. The cash prize was worth something like fifty quid and the trophies must have been worth about the same. When it came to the cup I think it was worth round about 200 quid, and the bracelet was worth somewhere around 100 to 150 quid too.

George and Lloyd Fogarty were successful in winning the V.R.D. Handicap with Typhoon ... The Trophy for this race was a travelling rug and an overnight bag, donated by Mr and Mrs C. Schultz, of Humbert River.

Hoofs and Horns, August 1953:56

We invited two nurses to come out from Darwin for one meeting. I think it was about the last held at VRD.[4] Their plane landed at the old airstrip about ten minutes before a race began, so I raced up and said: "Hullo! Are you the girls going to Humbert?"

"Yes, yes," they said.

"There's a race on now. Jump in quick and make the best of it. We might just get back in time. There's a horse starting there for each of you."

They didn't know what I was talking about and looked half-silly at me. I explained while we were going down: "I've brought over six horses and I'm giving you girls a horse each for the meeting." When we got there I still didn't know who was who, or what was what, so I said, "I've got the names of the horses that I'm starting in a hat. You just pick one." Each one took a name out. One girl got a horse called Potato and I think the other got one called Whiskey. I said, "They're your horses for the meeting. I've got them already picked out for the different races."

The girls' horses both ran several races each and one had two seconds and a win. Well you never saw anyone so excited in all your life! You'd've thought they'd won a Melbourne Cup, they were that excited about it.

The honours of the first day's racing went to Mr. C. Schultz with four firsts, three seconds and one third.

Hoofs and Horns, August 1953:56

Another race meeting I went to there, we brought over six or seven horses and out of the eight races we won seven. So help me Christ, I never in my life thought I'd hope somebody else's horse would win rather than mine. Coming down the straight in one race, there was one of my horses and a VRD horse running neck and neck, and I'll never forget it. My wife Hessie and I were both barracking and yelling for the *VRD* horse: "Let him come, I hope she wins, I hope she wins!" I think it did win too.

Oh, it was a terrible thing to go to a race meeting and clean it up — it really spoilt it for everyone else you know. It was so embarrassing we just went away and sat to one side. Talk about lucky though. Previous years we'd been over there and won one or two places, but we never ever won a damn race. Then this year we cleaned up the whole damn meeting!

> *Held a race meeting in June, very one-sided. Only interest in the social side. Decided to abandon all future meetings. Station inconvenienced for a week before and after the meeting.*
>
> George Lewis, VRD Station Report, 1963

By Christ, the fun and bloody games we had there. They were good old days and everyone used to look forward to the races, but they knocked that on the head at the last on account of the Aboriginals getting full citizen's rights. They were getting grog from everywhere and you could hear them down there in the riverbed fighting amongst themselves. George Lewis said, "It's gonna be black against white if we don't look out," so we had a bit of a meeting there and decided to give it away. We reckoned if there were going to be drunken rows or fights, well let it happen down at Timber Creek or one of those other places. We didn't want it round the station.

Races were held at other places besides VRD. There used to be a yearly race meeting at Timber Creek, or the Depot as it was known in those days, and as usual the stations came in from everywhere. In the early days you had to walk your race horses from place to place. They'd just trot along with your spare horses and when you got in the camp of a night-time you'd give them a feed of chaff from your packs. The races were only a hand-to-mouth turnout, but we always had a lot of fun.

This particular year one of the head drovers from VRD was there, a fella by the name of George Hunt, dead now of course. If Hunty had one or two beers he was a damn nuisance because he'd walk into your camp and be trying to tell you something, and he'd fall over the top of you. He'd get up, shake the dust off himself, and he still wouldn't see the trouble he was in and he'd babble away again.

Another fella there was Billy Linklater or Billy Miller as he was known out on the stations. Billy was a saddler and a cook — he followed anything on the station line — and he was a born poet, and a wit with it. He seemed to hate the boss drovers and all other bosses, although Billy forgot that he was a boss himself quite a bit in his younger days. Anyway, someone was taking a load of blacks back to Bullita Station in a utility truck and old Hunty had been making the usual insufferable nuisance of himself, so they grabbed him (so the story goes) and they put him on the truck and said, "Sit there you old bastard, and don't fall off because if you do we're not going to come back and pick you up." Billy Miller saw all this and put one and two together, and made up a piece of recitation about it. The last couple of lines went:

And seated amongst the Bullita gins
Was Drover Georgie Hunt![5]

God strewth! Someone recited this to Hunty a month or two later and did he hit the top rail over that! Whether it was right or wrong I don't know, but of course, Hunty wouldn't know himself. He was one of those, when he went out to it he flaked out to it properly. But God, didn't Hunty abuse old Billy Linklater for putting that in — and the more he kicked up a stink about it the further the poem spread!

6 VRD Outstations

Back in the early thirties the outstations on VRD were a thorough disgrace. Even the head stockmen were camping under bloody flies half the time. At odd places a bit of a shed was put up, but no cement on the floor or anything like that — the floors were antbed or some big flagstones. Some of the blokes that had a bit of a mule team dragged up flagstones, and they never got a thank-you for it either. The Gordon Creek outstation was twelve miles up the Wickham River from the main homestead and it was like that. All they had were a couple of tents rigged up and a couple of paperbark sheds.[1]

Frank Spencer was running the Gordon Creek camp for about eight or ten years and I had a lot to do with him there. (This was the same bloke who tried to rig one of the races at VRD that time.) He was only a little light fella with a reddish freckly complexion that got him the nick-name "The Red Hen" or "The Speckled Hen". Frank came from somewhere about Rockhampton and he got the wanderlust. He worked his way across Queensland and kept going further and further out, until he got to the Victoria River Downs and Wave Hill country. Eventually he got a job at VRD as a stockman in the Gordon Creek camp, and after he'd been there three or four years he got to be head stockman. I remember old Frank telling me that he'd never run a camp before and he reckoned he was lucky to strike Billy Schultz, because Billy was a great help to him running that camp. He went to the war with Jack Knox, the bloke who'd been running the Moolooloo camp, and later on they made Frank the overseer. He was on VRD for more than twenty years.[2]

I'd always attend the musters with Frank Spencer when

he was in charge of the Gordon Creek camp. As I saw Frank, he was a pretty good General running a stock camp. Nothing would trouble him to get things going, but there were some things he missed in learning about cattle. He never learned to work them steady, which was a fault with a hell of a lot of blokes in the Northern Territory. If they moved cattle they were only happy if they were galloping. They never considered that if cattle could be walked in they *should* be walked in, so as not to knock them around. Frank was like that. He'd have stockwhips going behind them and if they started trotting or cantering, well that never bothered him. It was all the same to Frank, but I can't stand knocking cattle around.

Now that they use helicopters it's even worse. Helicopters are all right for finding cattle, but definitely not for mustering — just forget about it. You run your cattle three and four miles, or five miles sometimes, and when they get to the yard their tongues are hanging out. And another thing too, a lot of the cows lose their calves. The calves get left behind and lie down. Usually a cow will come back to where it last gave its calf a suck, but the dingoes will follow up where the cattle have been running and find the calf and kill it straight away.

Frank's camp was the closest to VRD and sometimes they'd get an urgent request from the abattoir in Wyndham for say, 600 bullocks. I noticed it was always Frank they fell back on. Counting the two horse-tailers, the cook, and a boy with a three-horse dray pulled by mules, Frank had a good team of about twelve of fourteen boys. They'd send word out to Frank: "Do you think you could get 600 bullocks by such and such a day?" and Frank would send word back with the same boy. "Yes, I can get 'em." Nothing like that was a trouble to him because he knew where to go for them and except for that one fault that he had, working his stock too fast, he certainly got his bullocks.

From what I saw of him he could work Aboriginals too, although if it came to thumping them he wouldn't be in the show because he was too small. He knew how to work them, and the odd bad one he managed to get rid of. He'd tell him to have a holiday at VRD and he wouldn't give him a start when he came back. He got good work from them, but like all young fellas there were times that a few of them wanted straightening up.

Frank used to get me to attend the Gordon Creek muster because it meant there would be five or six extra hands. He'd say to me, "You bring all your Abos over and I'll look after 'em and feed 'em." So I used to go across and attend the muster and I'd bring all my cattle back — usually only about thirty or forty stragglers would have got over there you know.

Well, we'd muster and we'd bring all the cattle in to the camp, but we wouldn't be doing any branding. We were just after bullocks — we might have about 800 or 900 or 1000 on the camp. When you're riding through a mob that size, well that's a lot of cattle to be going through. You can clean them a lot better when you cut them into mobs of about 300. That way the bullocks are easy to find you see.

If you're drafting like that you just bring your bullocks slowly to the face of the camp, then whip at them and shoot them out quickly. But Frank would start his horse out jogging in the middle of the mob and before he got to the face of the camp there'd be other cattle running and jumping and getting out of the road, and by the time he got to the edge of the mob to shoot the one bullock out, there'd be about fifteen others going out with him. I could never make Frank out on that. Generally a good cattle man could see his own faults and would rectify them, but not Frank. And yet there were other good men there drafting too, and he should have seen the way they were handling their cattle, but no. That was really a bad fault with him.

When I attended the musters with Frank, there used to be an old yard at Police Hole on Gordon Creek that had great big coolibah posts about eight or nine feet high. The gateposts were still standing and there was a stub yard that I used to hold my stock while the muster was on. I'd cut out the Humbert cattle every day because VRD would be branding the next day. We'd cut out ten or fifteen or twenty, and hold them there. Eventually I'd go back to Humbert with about 120 or 140 head — cows and calves, steers and bullocks. The bullocks or any male cattle went into the bullock paddock.

One of the lousy things Frank Spencer did — all the stubs in that yard, he was pulling them down and using them for his camp fires. I couldn't see the sense in it because the next time I came along attending the muster with him, I'd have to repair it, and he'd even have to give me a hand to patch it up! Eventually most of the posts went, but there's still three or four there, standing just on the VRD side of the present Humbert–VRD boundary. Of course there were no fences between Humbert and VRD then.

I nearly got caught by a bull on one of those musters. I was coming across the run with a VRD stockman called Dan Macalvry. We had eight or ten cleanskins and we came across a young bull, so we tried to run him into the mob. If we couldn't get him into our mob, what I wanted to do was pull him over and tie him down — you always had bull straps to tie their back legs — cut his horns off, castrate him and let him go. At least he'd be a bullock then or a steer, and from then on, even if you didn't pick him up for another two or three years, at least he wasn't a bull any longer. He'd have lost that bull appearance.

Well he broke away and I took off after him. Dan was only a jackaroo and didn't know that much about stock, so he followed me. By jove, you talk about a cunning bull! Three times I started to leave my horse to throw him and he'd look back at me, and I'd stop where I was then — I wouldn't leave my horse. The fourth time, or it might have been the fifth, I left my horse and made my run, and just as I got about four or five feet from him he turned. I knew straight away I wasn't close enough to grab him by the tail and in the same instant I surveyed the situation for any handy trees — there wasn't a bloody tree within twenty yards!

I whipped round and I "off" to the nearest tree as hard as I could go, and the next minute I heard him coming up behind me, flat out. I got to the butt of a tree and whipped one arm around it to swing myself clear, but before I could do anything the bull drove one horn clean between my legs, and dragged me and threw me. He tossed his head and lifted me as though I was only a teaspoon, and shot me twenty five feet! I measured it after and it was a good twenty-five feet I shot through the air.

As I hit the ground my first thought was, "Now he's going to charge", and I was going to try to kick him in the forehead with both feet. Just how much assistance that would have been I don't know, but at least I would have been doing something. Well for once the jackaroo was in the right place, because the bull left me and charged at him. Of course I jumped up straight away and lost no time getting to a tree. He charged at the jackaroo a couple of times, but the jackaroo's horse must have been horned before that day because it "off" like a bat out of hell, with the bull right behind it. I thought at the time, "Too bad if the jackaroo falls off now, because then there'll be two of us on the ground." Dan hung on more by sheer fright than anything else and the bull left him and kept on going.

I had a bit of a pain in my left arm and my first reaction was, "I'm horned," but it wasn't that. When I'd put my arm around the tree and the bull dragged me off, the skin on my arm was severely grazed. If he'd have hit me with his sharp horns — he really had needle horns — he'd have put them right through me, but all he did was just put a little bit of a graze between my legs. It's those sorts of things that made you more careful the next time, so it was a good dose of medicine.

Another bloke I knew at Gordon Creek was Jack McDonald or "Big Mack". For a start he was the cook there and a darn good cook too. Later he became head stockman and everyone made a joke about how Alf Martin had to have a cook on VRD to run the stockcamp. He looked like a big plum pudding sitting on a horse, old Jack McDonald did. I didn't mind him. I found him to be a reasonable bloke, although a lot of others could never get on with him. One thing Jack could do was work with Aboriginals. When the war was on Big Mack decided to go, and he got down as far as Adelaide where I heard he was thrown out on account of having big feet. Somebody said, "No. Wrong. Wrong word, mate. He had *cold* feet."

Pigeon Hole is fortunate in having a team of natives that almost belong to one family, and a lot of young boys. This team can be depended to get their average every year while the team is held together.

George Lewis, VRD Station Report, 1967

Jack left the Gordon Creek camp and took over the Pigeon Hole, and it surprised everyone that he got his bullocks year after year. What carried Jack there was a couple of Pigeon Hole boys who were outstanding men. They were two brothers, Anzac and Hector. They both had families, and of course when their sons came along they were a big help in the camp too.

It was a wonderful camp, that bloody Pigeon Hole. They used to brand around 12,000 calves a year there. One time a manager came along and asked the head stockman, "How's your branding going Bob?"

"With this lot in the yard I'll be branding 12,000 and I hope to brand another 1500 to 2000 before Christmas."

Then the manager turned around and said: "Oh listen Bob. The Centre Camp down there is short of horses and you must have a hell of a lot of horses here if you can brand 12,000." That's what little brains that manager had. He said, "You don't need so many Bob. I want a plant of horses off you," and he took a bloody plant of horses off him. Of course they never branded another 1500 at Pigeon Hole then.

After Frank Spencer left Gordon Creek a bloody pom called Frank Reynolds took over. He'd been over at Legune under Jack Martin, but after he had Reynolds there for about six months Jack wanted to get rid of him, so he sent him over to VRD. He told VRD, "Reynolds can't get on with the bloody blacks here and he can't get along with the whites. You might be able to push him out of the road over there" — so they ended up giving him the Gordon Creek camp to run.

As soon as Reynolds started to brand up about the Gordon Creek camp, he'd always send over for us to come and get our cattle — that was his excuse, anyway. Well it meant that I had to take most of my men over there to get half a dozen head or whatever it was of my cattle, but I'd also be giving him a hand with branding and all that sort of thing, you see. We'd attend the musters with him and the way Roley Bowery and I used to help him at the Gordon Creek camp, getting his bullocks together! Of course he was more or less a raw recruit, but he was an ungrateful bastard, I'll say that about him.

I remember one year when Reynolds'd been there for some time, he was supposed to get so many bullocks to go into Wyndham and he was the best part of 200 short. I said to Roley, "What about it, what do you reckon? Do you think we ought to help the bastard or not?" Roley said, "Oh, it's up to you, but if you want to help him out, go back into that Kaiser Creek country. I think that would be the best place to pick them up."

Roley took two or three of my boys. I don't know whether he took Reynolds back with him or whether he only took some of Reynolds' stockboys, but he had a good plant anyway. There were about eight or nine men and they ended up getting about 300 bullocks. He was only supposed to pick up around about 200, but Reynolds, he just thought of it as a job that we should be doing for him; he was one of those ungrateful sort of buggers.

Tom Simpson was a cook at the Gordon Creek camp and he was a character all of his own. He was one that was on the overland telegraph line that came through from Alice Springs — he was well into his sixties or seventies when I knew him. Tom was very good to me and I got on very well with him. I think most people did. He wasn't anything outstanding as a cook, but he was a happy-go-lucky bloke. Actually there was a dash of Maori in him. When I attended the muster with him at Gordon Creek he used to help me along with any tucker that was left over there. He'd say, "Here, take this home with yer for yer blacks. We've got an overproduction here and our supplies will be out tomorrow or the next few days, and if you don't take it we'll only 'skie' it."

> *Wave Hill Cattle Stn. completely wrecked and several horses and cattle washed away. Travellers Drew, Bracey & Ashwin spent the night up a tree, and lost all their camp equipment; the M.C. lost a considerable amount of rations, goats and other private property. Heavy rain, river in heavy flood.*
>
> Wave Hill Police Journal, 12 February 1924

Old Tom was at Wave Hill at the time of the big flood they had there and he used to tell me about it. He said they built the place down in a bloody valley, a nice place for horses, but as regards a house, if a flood came look out! The flood they had that time covered the valley floor right through there.

After several years he left the Gordon Creek camp. I rode across to VRD one day to pick up my mail and when I got there, who should be camped on the river but old Tom. He was with old Harry Roper, another fella by the name of Wallace — Jack Wallace I think his name was — and Jack Noble. I think there might have been five in it and they were all going down to Tennant Creek to go prospecting.

Anyway, some of them being bits of old death adders, we knew they wouldn't get too far before they'd be rowing and wrangling with each other. I think before they got to Montejinni one bloke pulled out. Then they got up to Number 13 bore and two more pulled out. I think old Tom and Jack Noble carried on, but then Tom got a job at Newcastle Waters and Noble went down by himself.

I went back to Humbert and finished my muster, and about a month later I sent a mob of about 1100 steers off to Alice Springs. I went with them as far as Newcastle Waters and when we arrived there we heard about the rush that was on at Tennant Creek. My bullocks went on from there while I went back to Humbert and about four or five months later I got a letter from old Tom. He'd carried on from Newcastle Waters and gone right down to Oodnadatta — he came from that country. He was enquiring all about VRD and you could see he was a bit homesick about those parts. I really treasured that letter so I sat down then and wrote him four or five pages. About three months later the letter was returned unclaimed, and I heard he'd passed away.

Talking about Jack McDonald and old Tom Simpson being cooks reminds me of another couple of cooks I knew. Cooks were often old cantankerous fellas, too old for stockwork, and a lot of them hit the grog a bit. There was one I knew called Jack McGuigan, a little elderly fella and about the dirtiest cook you'd ever see. On account of being so dirty he couldn't get a job.

Old Jack hated the living sight of Alex McGuggan, the manager of Wave Hill, and McGuggan wasn't too partial to him. Old Jack reckoned, "That's not his bloody name, that's a 'nom-de-plume'. His name should be McGuigan." He'd always bring that up. He was camped on the Wickham River this particular time and he saw Alex McGuggan come along, so he said to him, "G'day Mr McGuigan."

"It's McGuggan, not McGuigan!"

"Oh, McGuggan is it?"

"Yeah."

Eventually they sorted their names out and McGuigan asked, "Have you got a job for a cook?"

McGuggan had no intention of giving him a job at all, but he wanted to tease him, so he said, "Let me see now, let me see. Cooking? Yes, yes. Oh that's right, cooking, yes, you're cooking—Aren't you the bloke that washes his feet in the corned beef bucket?"

Holy suffering Jesus! Yes — "Aren't you the bloke that washes his feet in the corned beef bucket?" McGuigan said to me later: "The bastard! Fancy saying that to me!"

Well he sat down there for about six or eight months on the Wickham River waiting for a job to come to him, but he didn't get one, so all of a sudden he said, "I'm off! I'm off!" No one believed him because he'd been there for so long, but he rolled his swag and went. "I'll go to Halls Creek and try me luck there," he said, and he was only there a fortnight when he died.

> *"Peggy" Wilkins, the one-legged stock rider, has died in Darwin. Despite his disability, he could take a turn at droving, and had been nearly forty years in the Territory.*
>
> Hoofs and Horns, June 1953:55

Peggy Wilkins was a cook too. Peg was a mighty man. I don't know how the hell he came to lose his leg, but even with it missing he worked as a cook or a boundary rider and always knocked around by himself with a packhorse and a couple of spare riding horses. The trouble was he was very unreliable. You could start him cooking and he'd be singing songs and what not, and in about a week's time he'd get up one morning and he'd put his time in, usually just as everyone was having breakfast or shifting camp, or something like that. I suppose you'd say he was a bit neurotic. He seemed to be happy to be on the move all the time, here today and gone tomorrow. Round and round — over to this station, he'd get a job, then on to another station, and another. At the last of it the stations wouldn't give him a job because the beggar wouldn't stay long enough.

Last but not least, there was Texas Jack and Blue Bob.[3] They used to knock around the bush together — Texas Jack was a cook, but Blue Bob was just an old bastard, fat as a match and about six feet tall. He was around Camooweal a lot and he'd get as full as a goog, and soon as he did he'd pick on the bloody police. He'd yell, "I'm Blue Bob the bastard and never been yarded! Yard me you khaki bastards, yard me!" and he'd let a snort out of him and kick out sideways. By Christ, they reckon he could snort too, you know!

The police those days didn't take much notice of anyone like that. If it was a bushman they'd shove him in for the bloody night and in the morning they'd give him a cup of tea and say, "Now listen, get out of this for Christ's sake, will yer." But as soon as Blue Bob got some more grog in him, back he'd come again, the bastard! He died in Camooweal I think, old Blue Bob. He might have been around in 1927 or '28, but he died soon after I got to the Territory.

Frank's plant turned up before dinner, Mick, Sailor, Monkey, Albert, Paddy & Marnaky.

Humbert River Station diary, 28 June 1929

They used to have a really good team of blacks there at Gordon Creek. One of them was Doug Campbell. When I went out there Doug was only about seven or eight. They put him in the Gordon Creek camp, him and another boy called Charcoal, and they used to have to lift them up to put them on a horse, but Christ, even then they could ride like buggery! Doug Campbell's father was a young jackaroo on Mount Sanford from what I heard, and later on he was a stock inspector round about Alice. By Christ, Doug's mother was a good sort of a gin. Her name was Maggie and she had the Roman features, no flat nose or anything, fairly tall, and good in the stockcamp.

Doug's blackfella father was old Sailor, the main man at Gordon Creek. One time Humbert was assisting the Gordon Creek mob to muster bullocks. Frank Reynolds went in to VRD and said, "I want Sailor," and they said, "Oh, he's sick" — he was a very crook man you know. Reynolds still went down to the bloody camp and rooted him out, and all the other blacks came up and said, "Oh, him sick fella that fella."

Christ, I wish I'd known at the time. I used to keep right away from VRD, but by Jesus I'd have bounced in there and taken old Sailor's part. Good old inoffensive blackfella! All he knew was work, work, work, you know. That bloody Reynolds worked him right into the ground because he had to get his bullocks, and the only one he could depend on was Sailor. Reynolds couldn't do anything himself he was that bloody useless, the bastard.

Another boy called Albert was pretty good too, but the dandy of the lot was Greasy Bill. Greasy looked after the packs and all the packhorses. He had those bloody mules tied up before you were finished your breakfast, and he had the pack saddles on them, and he'd weigh the bags — Christ, he worked damn quick! I don't know what the hell happened to him. I went to Queensland on holidays and I heard he'd died while I was away.[4] Then there was Paddy. He wanted a belt under the bloody ear that fella. He had the wind up Spencer too, but Old Albert, Sailor, and Greasy Bill were the main Gordon Creek men, and Doug when he got to about nineteen or twenty, him and Charcoal. It was a bloody good team there then — I'm going back about forty or fifty years now.

I had Doug working with me for a while on Humbert. Old Alf Martin didn't like half-castes and one day he said, "Look, why don't you take that bastard over there?"

I said, "Spencer won't like it and I don't want any strife with the Gordon Creek camp."

He said, "Take the bastard out of here."

Of course a lot blokes reckoned that the half-castes would hang around and get all the information from the whites, and then they'd reckon they're "big fella men" and they'd go down and put it over the blacks, you see. And vice versa — they'd hear what's going on with the blacks and they'd come up and start telling the head stockman. It could be right too. When Doug Campbell and Charcoal were young fellas they were two good men, but then when they got in amongst the whites they turned out unreliable bastards. They'd upset the apple cart when everything was going along nice and smooth in your stock camp.

> *Received message from Constable Fitzgerald to effect that an Aboriginal named Humbert Tommy had assualted [sic] him & his private boy George.*
>
> Timber Creek Police Journal, 15 August 1937

It was Frank Reynolds who set off all the strife when Humbert Tommy got into a fight with the police and ended up spearing a tracker. Reynolds came on a couple of dead weaners, and of course he always had his knife into Humbert: "Those Humbert boys are coming over here and spearing cattle," and he made a complaint to VRD. It must have been Alf Martin who sent word to the police at Timber Creek.

The police sent out a bloke called Fitzgerald who had no experience — just a raw recruit. "Fitzi" came down and said, "Who was it?"

"A Humbert boy — Humbert Tommy."

"Oh, that's one of Charlie Schultz's main boys."

"Yes, he's spearing the cattle out there."

Tommy was about one of the best boys I ever had. I forget the name of his mother, but that Gordon who was shot up there at Light Creek was his father, and Humbert Jack and Humbert Tommy were half-brothers. They were old boys in the stockcamp when I first went out to Humbert. Tommy was a very quiet, serious sort of a bloke and very surly. Humbert Jack was taller than him. He'd be about six foot and he was a quiet sort of a bloke too. Those old boys, I got on really well with them and never ever had any trouble. They knew there was never any backing and filling with me — a spade was a spade and once they knew that it was all right.

Tommy was camped with his gins and a couple of dogs up at the head of what they call Steep Creek, and he was killing bloody cattle there, catching weaners with his dogs. He'd spear them, or his dogs would grab them and pull them down, and he'd just go and knock them on the head. Oh, it was like getting money from home with him! When I heard about it later I was surprised he'd been doing that. I don't know why he wasn't working for me at that particular time — he'd walked off camp once or twice so he might have run away.

When Fitzi came, Roley Bowery was at VRD and he was roped into the set-up. Fitzi said to him, "You'd best come out with me while I get this bloke", so they went out then to raid the camp. Later on Roley told me all about it. I've got a feeling they were only going to give Tommy a kick in the pants and send him back to Humbert with Roley. Otherwise I can't understand Roley going out with them, because generally we kept away and let the police do their own work.

> *There was a number of dogs in the camp, and as aboriginals were only allowed to have two he [Fitzgerald] shot one and maimed another. They both belonged to Humbert Tommy.*
>
> The Northern Standard, 26 April, 1938

Anyway, they rounded up Humbert Tommy — he didn't run away — and Fitzi said to him, "What are you doing here?" or some damn thing. According to Roley, Fitzi went about it the wrong way. Roley said Fitzi was talking to Humbert Tommy in an ordinary way you know, and he said, "You've been spearing cattle here." Tommy didn't answer him and Fitzi said, "Well, you'll have to get rid of those two dogs." I think Tommy agreed for Fitzi to shoot them if he wanted.

Fitzgerald had a tracker called Kelly with him.[5] Kelly shot more bloody Abos than enough, that bloke. He's dead now, but by hell I tell you what, if there were any outlaws the police wanted, that Kelly'd go out. He was frightened of nothing. I always remember Gordon Stott saying to me, he said, "He's the only nigger I've had working with me, Charlie, that I'd say has got guts." Gordon said there might be twenty of them there and if he wanted anyone he'd give Kelly a pair of handcuffs. Kelly'd go into them and be putting handcuffs on this one and on that one, and kicking blokes out of the road, and he'd stop at nothing. And the same way with the whitefellas, he was frightened of none of them!

Accused had several stonehead spears and a shovel-nosed spear ... He said "I'm no more — myall black: — I won't sit down" ... A scuffle followed ... and [Fitzgerald] struck Tommy a blow on the head with his revolver.

The Northern Standard, 26 April 1938

Tommy had some spears in his hand and Fitzi said, "Put those bloody spears down," and Tommy said, "I'm not a bloody myall," and shook the spears at Fitzi. I don't know whether Fitzi clouted him with his revolver or if he said, "I'm putting you under arrest", but Roley said there was a great big round rock there, and the next minute Fitzgerald yelled and squealed like a stuck pig and came straight over the top of it.

I think Roley was on the other side of the rock looking after a couple of Tommy's gins so they wouldn't run away. "Christ!" he told me, "If anyone did their bloody block, Fitzi did. I honestly thought he was bloody well speared, the way he yelled and squealed. I let the gins go and raced around and saw Humbert Tommy coming, and I went within a bloody fraction of giving him a bullet as he passed by."

The next minute Tracker Kelly raced up and Tommy turned round and let fly at him with a spear. Kelly threw himself on the ground and the spear ran right along his ribs. He pulled the spear out and threw it back at Tommy. It was only thrown from the hand, it wasn't in a woomera or anything. Roley saw all this happen. Humbert Tommy dodged the spear and then ran up the side of a hill towards a lot of rocks near the top. Kelly was after him, just one step behind him all the time according to Roley, and just as he got up near the top, Kelly collapsed. Humbert Tommy kept going and took off.

Well they went up to help Kelly then. They knew he'd been hit with the spear because they saw him pull it out — it was the loss of blood made him collapse. They got him back to the station and there was a big hullabaloo: "A policeman nearly got speared out here and one of the trackers got speared," and so on. Of course, they should never have sent an inexperienced man down in the first place.

More people came down from Timber Creek and made a great song and dance about it all, but before anything else happened, Tommy gave himself up. The police made themselves out to be big men who went out and arrested Tommy, and the "heroes" took him down to Darwin and put him in clink.

It was Tas Fitzer that got Humbert Tommy out of trouble. Tommy only spent about a year in jail before they sent him back out to VRD. He was getting a bit too old for stockwork then so he spent most of his time sitting down in the blacks' camp there. Years later Tommy borrowed a .22 rifle from a bloke called Charcoal and shot himself with it. I don't know why he did it — he was just about the only blackfella I ever heard of who committed suicide.[6]

We seem to have a couple of serious cases of blackfellow trouble lately ... [one] incident occurred in that corner of the country where the blackfellow has carried and earned a bad reputation for many years — the Montejinnie area of V.R.D.

Hoofs and Horns, February 1956:6

About the same time that Humbert Tommy speared Tracker Kelly, Snowy Shaw got hit with a boomerang. There was a little tribe of blackfellas around Montejinni — I forget what they used to call themselves, but by Christ they were a bad bloody mob. Snowy was head stockman there at the time and he was what we used to refer to as a "good blackfella man". As soon as he got a nip or two in, he'd go down and start sitting with the blacks and talking to them, and then if any of them gave him back-slack he'd haul off and belt them.

Snowy did this again one time and he got into a scuffle with old Kaiser Bill. Oh Jesus Christ! Kaiser hit him with a boomerang and stuck the bloody thing into his chest. They were terribly frightened that Snowy'd get pleurisy from it. They reckon when he breathed, blood was spurting out you know. Anyway, a fortnight or so later he was back riding horses.

> *Stating that he was satisfied there had been sufficient provocation, Mr N.C. Bell, S.M. ... found that Kaiser ... was not guilty of unlawfully assaulting George Shaw, head stockman of Montejinni station, on July 19th, and he dismissed the case.*
>
> The Northern Standard, 9 August 1938

Snowy told me about it himself and showed me the scar about an inch and a half long on his chest. He said, "Charlie, I don't know. I only get like that when I have a few."

I said, "It serves you bloody well right Snowy. You *will* sit down and play with them and talk with them. They're not educated enough, and they regard you as a boss and someone that's over them, not level with them. That'll only happen in times to come."

7 The Upper Wickham

> ... the Western areas are chiefly rough broken sandstone with spurs and ridges and rough gullies of little grazing value. A large proportion of the sandstone country is inaccessible and useless for working stock.
>
> *Investigation of Pastoral Leases — Northern Territory, 1934*

When I first went to Humbert that upper Wickham country belonged to VRD. It's very rough country and VRD never bothered to even have a look at it, so I used to muster it once or twice a year. VRD didn't know the first thing about that. As a matter of fact I didn't know whether it was on Humbert or whether it was on Limbunya or VRD or Kildurk — it was a sort of Never Never country.

Knowing where the boundary was used to be a problem for all the stations in those days so it was a matter of give and take. You'd muster different areas depending on where your homestead or your stockcamp was and what the country was like as well. The stations weren't fenced and they didn't take too much notice of the boundary on the ground — they looked at the way the country lay. It's no good this "dog in the manger" stunt.

> *The readjustment of their boundaries — or what they have hitherto looked upon as their boundaries — has come as a great and a not too pleasant surprise to a few. It is found, for instance, that the Victoria River Downs boundary runs 25 miles further to the southward and 10 miles further to the eastward than was supposed by some of the settlers ...*
>
> *The Northern Territory Times, 23 March 1906*

Before I ever got to that country there'd been a dispute

about the boundary between VRD and Wave Hill. Somebody on VRD got the idea that the southern boundary of the station might be further south than was thought, so they asked for a surveyor to come in. Wave Hill must have had an idea that they'd lose out, because they didn't want any bloody boundaries surveyed, or fenced either! It turned out that Wave Hill was encroaching on VRD country to the tune of twenty-seven miles, but even after it was proved that it belonged to VRD, Wave Hill came in there three or four times on the excuse that they were getting their cattle.

Those big stations didn't like locking horns with each other, but Tom Graham said, "Well why the hell didn't you give me notice you were coming down here?" Tom started up a camp at Mount Sanford to keep the Wave Hill mob out of it, but he died before he could get a fence put up. When Alf Martin became manager of VRD he got Wally Dowling to put up the boundary fence between the two stations.

On the western side of VRD, ranges formed the boundary and it was never fenced off. You can't get in there for a start and you'd have to see it to believe it — but just because the upper Wickham was VRD country they didn't worry about me going in there, and I was the same — if VRD wanted to go and get their bullocks in parts of Humbert that I couldn't muster, well I never worried about that.

For instance, there was a part of Humbert that went south-east across the Wickham River, but from the position it was in I couldn't get into it. Mount Sanford could get into it much easier so they used to muster it. Elmore Lewis worked at Mount Sanford and he asked me, "Why don't you go in there?"

I said, "What's the country like?"

"There's a bit of good country in there," he said, "but it'll take getting into from your side. We go in there because we get a certain amount of bullocks from VRD."

In other words, it was more convenient for me to muster their country on the north side of the river and for them to muster part of Humbert on the south side. Of course we never let any cleanskins go — a cow in the hand's worth two in the bush. VRD got them or I got them, and that was it — we just branded them. We'd work in with each other, but when Frank Reynolds was running the Gordon Creek camp you couldn't work in with him. He was one of those, "No, oh no. I couldn't do that or couldn't do this" fellas. Well we used to just carry on and ignore him.

It suited me to go up the Wickham and muster down. I'd only take a small plant of about eight or nine men, but VRD always had a big plant of horses and stockmen. I think they had about fourteen men in a stockcamp and each man had four horses for a start. Then there were six or eight spare horses in case of any going lame, and besides riding horses they had about twelve packhorses and eight night horses. With a big camp like that they had to be continually on the move, but I mustered in a different way from VRD — I took my time. VRD reckoned it just wasn't worth their trouble to go into the really rough areas, because the returns in cattle mustered never covered the effort or costs involved for them. They had more good country than they could handle where they could brand five or six times the amount of cattle they could get up there, and they didn't have to chase them.

They were all wild cattle in the back country of course, and it was all rough ranges and gorges. By God you know, there were some cattle up there in that Wickham country! I think they came across from Mount Sanford. You'd see bullocks in mobs of fifteen, twenty or thirty, and a lot of cleanskins in them too. They'd be up to ten and twelve years old with great long horns on them. Roley Bowery and I used to muster all through there.

I put up wire yards right through the Upper Wickham, which made things more convenient for mustering. Instead of watching your cattle, which can be very tiresome, we'd shove them in a yard of a night-time. Then of a daytime I'd take out about eighty or ninety for coachers and leave two boys to mind all the rest. They'd just feed them around the camp. Another thing, if you were driving them around all the time they'd get that damn footsore they wouldn't be able to walk.

I wouldn't tell VRD I was going to muster there, just in case they did decide to make the effort to come in before me and muster the bullocks — and take the cleanskins too, naturally. I'd just go in and if I saw a mob of cattle I'd start them up. The bullocks would be in the lead and you'd whip in and let the bullocks go to hell, but hang to your cleanskins as much as you could. Then if you got some bullocks into the coacher mob they'd be racing around wild as hawks, looking around ready to make a break. All of a sudden two or three bullocks would come to the edge. Right! Let them go, but hang to your cleanskins. That was one of the main things that I drilled into my stockboys: "Never mind about the bullocks. If you're having trouble with your cattle and one breaks away, let that one go to hell. More better for you to hang to the lot than lose another five or six or eight or ten" — better to lose the one than to lose the lot.

I used to muster up there every year or every second year, and pick up around about 200 head of cattle, which suited a battler on a small place like Humbert. It wouldn't suit the likes of VRD with their big herds — they could get five or six times the number of branders much closer to the station you see. VRD couldn't control their cattle in the easy country, let alone in the rough areas.

When you're on a station there's no mistake, there's different incidents that you see. I always remember a maiden-spayed cow I found when I was riding along Police Creek one day, about eighteen to twenty miles from the station. A maiden-spayed cow is one that's spayed at about twelve months. Usually you catch up with them the following year and cut their horns off, but this one went way out back somewhere and I never caught up with her till she was about seven or eight years old. Her horns weren't taken off.

We were camped at what we called Top Humbert Yard and I took a ride out in the afternoon to see what cattle were around. I saw a beast standing in the bed of the creek, and when I came closer I could see from the way she was standing with her head cocked to one side that she was listening. I thought: "That damn thing's blind!"

I drew up on top of the bank and looked down at her, and I could see the horns had grown anything up to three inches into both eyes — they'd filled the eye sockets. She was in reasonably good condition, but was on water and too frightened to leave. That was one of the few times I never had a rifle or revolver with me. It was too far to go back to camp for a gun and come back again that same day, and the next day we were going out mustering in a different direction. I couldn't very well hold up the plant for a day, so I never got back to put her out of her misery.

Sometimes cattle will put their heads in the fork of a tree to scratch their neck. Up the Wickham River there a heifer who did that got her head in, but couldn't pull it out. When I found her she'd been dead for some time, but you could see where the damn dingoes had come along and eaten the back of both hind legs, and the thighs. They'd nearly had her turned inside-out too; they'd opened her up on the side of the belly and her intestines had come out.

When grass gets scarce towards the end of the year a lot of cattle end up fairly low in condition, more or less starving. Often they get in on the banks of the river eating green pandanus, and sometimes they'll get in that close to the edge of the bank that they'll slip and fall head-over-turkey into the water. You'll find them lying on their backs with their legs sticking up in the air. Others will fall against a tree or a root and if they can't struggle out and get to the bottom, they'll die there.

There was an incident up at Halls Pocket — I'll never forget this day. Towards the end of the year I came along there and across the other side of a waterhole I could see a three-year-old heifer in the water, with her first calf up on the bank. I thought: "That bloody thing can't get up." I tied up my horse and stripped. Then I crossed over to the Halls Pocket side and went up to this cow and sized up the situation. She was a beautiful little red heifer and her little calf, about three or four days old, was lying on the bank there. I thought, "How the hell is it that a dingo hasn't caught you?"

What had happened was that the cow had fallen in the water among the branches of a tree that had come down, and every time she'd try to get up the bank there was one branch that caught her across the neck. She'd tried to get up that many times that she had a patch of hair worn off her neck. I managed to get her out in the water, but she was too far gone. I ended up having to shoot her, and the calf as well. And she never tried to charge once. Generally when they get down like that they bawl and put on a hell of a bloody turn.

By Christ I felt that, having to shoot her. She was so quiet and there was that little calf just lying down there looking up, wondering why her mother wouldn't get out I suppose. I persevered with her for a bloody hour there and at last I said, "It's no good." There were nearly always dingoes around there — whenever I rode that way I always had my rifle ready to get a crack at them. Well I had to get my rifle and bloody-well blow her over.

Later I thought, "What if anything had happened to me?" I was always frightened of getting a buster out there. A man could go one or two days before the bloody blacks back at camp would think, "Oh, something must've happened to that *Sharlie*, him not come home." Poor old *Sharlie*'d be getting cooked in the bloody sun!

---

1 yard 40 yards by 30 yards on head of Cusack Creek. (Made of stone and known as Stone Yard. (Built 1936) Value: £50.0.0.

Charlie Schultz, 8 August 1944

Well I wanted a yard somewhere in that upper Wickham

country, but it was mostly limestone and you couldn't get a decent bloodwood posts to put up a yard. I was always on the lookout for timber for a yard and if I could find posts within a mile I'd snig them in with a bronco mule, one at a time. It took a long time but eventually you got them there. You carried a crowbar and shovel into that country over your shoulder, dug your holes and put your posts in. Then you'd drill holes for the wire with an old brace and bit. Each yard used to have ninety to a hundred posts, counting the wings, and I had to do all the drilling work too because the Abos could break the bit, you see. You can break a bit quite easily.

Up Cusack Creek one day back in about 1936, I was riding around and I saw a little bit of a pocket in the side of a hill. This pocket had a low stone bank at the bottom enclosing a bit of flat ground, with a bit of a creek running through it. I had it in the back of my mind, "I wonder could I make some sort of a stone yard?" There were plenty of flat slabs of red sandstone in the area and I got carried away with the idea. I thought, "In case I do come back to put it up, I'll have a look around where I'm going to put it. I walked around and I thought, "I'll go from this tree to that tree there" — I suppose about eight or nine yards.

I had a couple of black boys with me so I said to them, "Might be we make'em yard here someday but I want to see how long to put'em stone here." I used to explain to them and tell them all that sort of thing, whereas a lot of blokes would just say, "Go on, put a stone down there," and half the time the blackfella wouldn't know what the hell they were doing or what it was for. I said, "We're going from here, straight to that tree there. I want'em straight line now and we make a wall for a yard."

They started getting slabs and I put them in while they were bringing up more. They didn't have to go that far to get them, about twenty yards or so. I kept on putting the stones down and soon realised I had to make the walls perpendicular inside. Otherwise, if there were any steps, the cattle would be jumping out. It took me an hour to do about ten yards of wall about three feet high, and I was really carried away.

I couldn't get home quick enough then. I grabbed all the available labour I could get and I said, "We're going up there now to make a yard out of stone." You should've seen the looks on their bloody faces — going to make it out of stone! I said, "We'll get a big mob of stone together and make'em yard." They had no idea what I meant, but I suppose they thought, "Well we'll go up and see what the old fella means." There were about six stockboys and I said to a couple of them, "Do you want to take your lubras or do you want to leave them home?" They saw their lubras and some of them wanted to come — it was a sort of a walkabout or holiday for them. There were horses for them to ride of course, and they said, "Oh, we want'em saddle." I said, "Yes, all right. There's saddles over there for yer."

I had three or four lubras who could ride pretty well you know, and we used to have them tailing cattle. They weren't real fast riders, so I'd always put a boy with them in case something went wrong, but oh Jesus, old Daisy and Hillary weren't too bad, and Molly wasn't too bad either! I had a good team of gins there.

We got stuck into it and finished the yard in five or six days. For the gate we managed to find and put in two posts. I don't know where I went for the gate rails, but I got them from somewhere. This yard turned out to be a real cattle-trap too. They'd walk down from the hills and jump down the stone bank, but they couldn't jump out again. They were down in a sort of a hollow.

I went up to Stone Yard one time with a mob of coachers to start mustering. We got in that night and yarded the coachers — I think they were just as silly as we were, the way they were looking around at the stone, being in a yard you know. I put the rails across the gate and went down and camped just below the yard on a spring in Cusack Creek — there's springs all through that country. A lot of the cattle in that Upper Wickham were unbranded and wild as hawks, and next morning when we went up to the yard, so help me Christ, there were about eleven bulls in it! They'd walked down the pocket and jumped in, and the buggers couldn't get out. God, did they fizz around the bloody yard and run up and down, but they couldn't jump out. I had a .303 rifle and wanted to shoot the buggers, but I didn't like shooting them because I wanted to muster that day, and the echo of the rifle would carry for a long way and frighten the cattle. There were two or three real wild buggers and as they came out, I waited until they broke and — *bang* — hit them in the guts. They ran away about a quarter of a mile and pulled up under a tree and died there. I did well out of my first muster. When I left for home six days later I'd picked up about 360 to 400 cleanskins — all ages, from calves up to bulls three or four years old.

> *The manager of Auvergne stated to me that <u>thousands</u> of bulls would eventually be shot on adjoining stations in the Territory. These bulls should have been prime bullocks and, as can readily be understood, are a menace to the breeding cattle, heifers often being killed by their maltreatment.*
>
> F.J.S. Wise, 15 August 1929

We shot a lot of cleanskin bulls in the early days. You didn't get any money for them then. We didn't try to catch them to castrate them because they were as wild as hawks and you were only knocking your horses around running the buggers down. The easiest way out of it was just to give them a lump of lead — you didn't even have to get your horse out of a gallop.

After the Second World War the Americans began to buy bulls for hamburger meat[1] and there were very few ever got away from us then. I always remember when the buyers first came out to Humbert, they said, "Have you got any bulls, any old bulls? Don't shoot 'em. We'll give you £8 a head for 'em." Actually that was as good a price as we were getting for a bullock. I didn't come in on the deal straight away. I thought, "If these blokes can come out and offer you £8 a head for your bulls, they'll be worth more than that." I knew they'd be keeping their prices down. Others were rushing in, but I hung fire and about three or four months later I got £10 a head for mine.

To catch the bulls we got after them on a horse and pulled them down, threw them and tied their back legs. We had pruning saws — that's a half-moon saw you know — and we'd whip their horns off and then bring the coacher mob along and pick them up with it. We'd take their horns right off close to the head, but wouldn't castrate them — that would bring on bleeding and you'd only drive them on about half a mile before the loss of blood would weaken them. You couldn't drive them then — they'd just bail up, jack up properly, and all they'd think about was charging you.

Often a bull used to break away and try to go bush, but you'd meet him with your horse and shoulder him and push him round till you got him back in the mob. We'd bash him over the head with a stick too, a green stick of course — a dry stick would break. We'd get him back into the mob and he'd go to charge out again and everyone would meet him on their horse, come in and yell at him. After about the fourth or fifth time you kept out of his road then, you didn't tease him. When he went back in the coacher mob you stood off him. He might go to charge, but usually he'd only come out about ten or twelve yards and then pull up dead in his tracks, and shake his head. We didn't go up near him and annoy him. All of a sudden he'd back away, and turn and run back into the coacher mob. You knew you had him then. Those same bulls, the next day they were practically on the tail of the mob and you couldn't get quieter cattle out of them.

Well we'd take them home and then we'd castrate, brand and earmark them, and put them in the bullock paddock. We generally kept them on the station the first year while they were still getting over the castrating, and over the knocking around they got when we first picked them up. I remember one year I went across to Bullita and we ended up bringing back to Humbert seventy bulls that we'd thrown, tied and dehorned in the bush. Some of them were big fellas up to four or five years old.

Not long after I built that Stone Yard, Alf Martin, Frank Spencer and two or three blackfellas were cruising around the Upper Wickham. They weren't doing any mustering — they just wanted to have a look at the country. Alf Martin had never been up there before — those days there were no planes of course, no helicopters or anything like that. There were rumours about big plains on the Upper Wickham. Well there were plains too, but they were only third-rate plains, not like that downs country around Mount Sanford or Pigeon Hole. Far from it. There's no Mitchell or Flinders grass in there for a start.

They camped near that stone yard, right at the spring. The yard was back about 150 yards I suppose. Pockets in that country up there have a wall of rock right around, and they merely looked up — no doubt they saw the stones, but from a distance they thought it was natural.

After they got back Frank Spencer said to me, "We couldn't find any yard or see where you camped or watched your cattle."

I said, "Oh, I've got a bit of a yard there."

"We thought you might have, but where the hell is it? And where would you get enough timber?"

"Well actually I made it out of stone — it's a stone yard!"

Later VRD let the Upper Wickham country go. I was in Darwin one day in the late forties, and I was speaking to a Mr Picket from the Lands Department. We talked about one thing and another and then he said, "Charlie, VRD is throwing open the up-river block. Why don't you take it on?"

I said, "Well I always muster that block for the simple reason that VRD never goes near it."

"I know that, but now they've decided to let it go and the likes of you being their neighbour, I suppose it's the old saying, 'better the devil you know than the one you don't.' You wouldn't have any trouble getting it."

The original Humbert block was only 579 square miles and a bit on the small side, so I was interested: "How much a mile are they charging on it?"

"About two shillings a square mile."

I said, "Right. I'll take it up."

So right there and then I took it up and added it to the Humbert block.[2]

8 "To Queensland, Droving"

In 1930 I mustered 990 bullocks. I'd never put a big mob together before and by gee that knocked me around. If I'd had the experience then that I had a couple of years later I probably would've got away about 1300 to 1400, or even 1500. Of course Billy Schultz hadn't been sending any bullocks away at all. By Christ, the place was in a mess! As well as mustering by day I was up half the night watching them — that's what takes it out of you. I had no yard in the bullock paddock. A lot of them got out, but there were still a lot in there that didn't get away.

The first mob of bullocks I sent away I handed over to Jack Morck to take to Alice Springs. There's a billabong on VRD named after Jack. According to old Jack, in the early days he and his brother Fred brought bullock teams up from Alice Springs to Katherine, and then went out to Wave Hill. He said some of their bullocks went lame and they had to pull up on that billabong for a couple of days to shoe them. They put up a V-yard there, dragged the bullocks up, pulled their leg back and tacked on shoes. They'd had to do the same before they got to VRD, at what we called the Cueing Pen there on Delamere Station. A couple of those posts were still standing on Delamere when I first went out to Humbert.

I started Jack Morck away with 975 head and a station plant. Of course with freshly mustered cattle they're restless all night, so we had to put six men on watch. We only had two watches, each half the night long, and that's very hard on you because you're working all day as well. Once they'd settled down I took my plant back to Humbert, but before Jack got the cattle to Alice Springs he lost about 230 or 240 head, just through carelessness. They reckon at daylight there were often bullocks scattered everywhere.

With what I got for those bullocks there might have been just enough to pay the interest to the banks that year. It's that long back I forget now. The next year I started about 1200 cattle away with a fella by the name of Arthur Grace. Jack Morck was offerd this mob first, but he knocked it back. He reckoned he was getting too old and was a bit disgusted with losing so many of the first lot. He was a man well in his seventies then — too bloody old for droving! I gave Arthur a start as far as the Wickham River and told him where to camp on the other side. Then I rode back ten miles to where I had about 250 bullocks, real old pikers and shelly blokes that I was taking into Wyndham. After I left him, Arthur started to cross his mob over the Wickham River, right at the Humbert junction. When you cross them over like that you want to push them across. Cattle have their own beat or home run and usually stick to one side of a river. He just let the lead cattle run down and have a drink, and then they turned back onto him.

He lost about 400 that night — luckily they were still on the place when it happened. He mustered the next day to make up the numbers, and what the stupid bloody fool did was to turn round and bloody-well grab cows and calves. He put the cows and calves with the other bullocks and steers he already had, but some of the calves in the mob were only two or three months old. Well in no time the steers were walking the heads off the cows and calves! It meant that when he got to Newcastle Waters or further down the line, the cows and calves were holding up the steers.

Grace knew he couldn't get in touch with me because I was on the road into Wyndham, but he knew that Cunningham was standing to us and if he had any problems he could wire him. Cunningham suggested that he split the cattle into two mobs, which is what Grace did. Two men had to bring the cows and calves along at a slower pace, which meant more money going out to pay for extra hands. And instead of trying to keep expenses down, at the Newcastle Waters store Grace bought up what he could. I forget how many dozen eggs he bought and there was booze into the bargain as well. When they eventually got to Alice the cows were sold for a hell of a giveaway price. Later on I heard that early one morning, when everyone else was still in their swag, Gracie had cut a wing — about 200 to 250 — and sold them for £1 a head. One of his stockmen woke me up to it, a bloke by the name of Eddie Kemp. At the time I wouldn't believe Eddie because I knew that he and Grace were always at each other, and I didn't think Gracie would come at such a thing.

After I'd given Arthur Grace his start with my cattle, I rode back to the other mob and started them for Wyndham. It was a twenty-eight day trip and it wasn't till I arrived that I got a telegram from Grace saying, "Lost two hundred" or some bloody thing. I thought, "God suffering Christ! What's gone wrong now?" I think I'd lost three old bullocks myself before I got to Wyndham and I was disgusted about *that*. Needless to say I had some sleepless nights then.

I got £1.5.0 for the bullocks I took to Wyndham, but I got the money in three different payments. For a start I think I got ten shillings a head on them which gave us something to carry on with. About a week or a fortnight later I got another ten bob a head. At the end of the year I got the last payment, about five shillings a head for the hoofs and horns. Part of the money went to pay old Matt Wilson for the stores he'd given me on credit the year before. Well we didn't have enough to pay the interest on the place that year, yet there was over 1400 head of cattle went off the place.

Arthur Grace was using a plant from Humbert, so after he delivered his mob he bought the plant back and put in Christmas with me. I had three well-bred mares I wanted to get in foal, so he said: "I'll take 'em across to Wave Hill for yer if yer like, and get 'em in foal there." Wave Hill had about five or six stallions, good quality southern stallions.

When the day came to go over, I said, "You can see old Owen Cummins up there. He's in charge of all the stallions, but go down and see the manager and tell him the mares are from me. If he says 'no', well that's all there is to it. Just bring the mares back home again."

When Grace got there, old Owen told him where the stallions were. "Take your mares down and let 'em stop for a couple of days and take pot luck with 'em," he said. Well Arthur did this apparently, and then made tracks for home. When he got to within about a mile and a half of Humbert, on a good little spot on a creek there, he left three or four of his horses, but my brood mares he brought home. He told me, "I've got a surprise for yer up Peter Creek."

I wondered what the hell the surprise could be. "Ah, you'll see when you get up there," he said, "Get your horse and after lunch we'll take a ride up there and have a look." It really had me jiggered you know, this "surprise" business. I thought maybe there was a whitefella there or something. I really didn't know what to expect.

When we got there I saw these four or five horses, three or four of his own and a stallion. He said, "What do you think of the stallion?" I looked, and I said, "Jesus! Where'd you get him from?" thinking he must have bought him.

He said, "I lifted him."

"You what! Where the hell are you going to put him?"

"I thought I'd give him to you."

"Well I don't want it if it's hot!" I said.

I think Arthur was hoping to get my next mob for delivery, so he'd brought me the stallion to try and make things good about "losing" the 200 head, but you only had to use your common sense. I was the only private bloke in the locality. All the rest were company stations and those big companies wouldn't bother pinching off each other. I'd be under suspicion immediately. There was a police station right there at Wave Hill — they could've reported the stallion missing as soon as they found it gone. We were hoping that old Cummins wouldn't miss him for a couple of days to give Arthur time to get it away. Arthur didn't like my style at all about that, but I insisted. "Look. Take the bastard to hell out of here Arthur; I don't want him at all," I told him.

Arthur camped with me that night and next day he saddled up and away he went. He didn't tell me where he was going and I didn't ask. My one ambition was to get the stallion off the place.

About a week or so later a gin sang out, "Packhorse bin come up, boss." I looked, and it was Grace coming in along Peter Creek. He didn't say what he'd done with the stallion and I didn't ask, but I thought he must have either taken it back to Wave Hill or else let him go up the Wickham River.

It turned out that he didn't take it back to Wave Hill because soon word was around that a Wave Hill stallion had been lifted. Frank Spencer at Gordon Creek mentioned it and also Harry Rowlands at Pigeon Hole. I reckoned then that Grace must have bushed him out on the Upper Wickham. It's all broken ranges and gorges up there with lots of springs — wild country that no one ever went into. It'd be difficult for a horse to find its way back.

Well I know I got the blame for pinching that stallion. Over the years different ones used to drop hints to me: "What'd you do with that Wave Hill stallion?" or, "Have you got any foals by that Wave Hill stallion?" I used to tell them, "I've got no Wave Hill stallion on Humbert" — but they'd expect me to say that anyway.

Years after when I was in a plane flying over the head of the Wickham, I saw about fifteen or sixteen horses there and I couldn't help but wonder whether these could have been some of the progeny of that stallion.

The next lot of cattle were taken by Roley Bowery. To help Roley make a good delivery I took my station boys and went as far as Newcastle Waters with him. Going across the Murranji Track out past Montejinni, the bullocks rushed. They rushed just about every night from Number 13 bore right across to Newcastle Waters.

God strewth! That was my first experience of cattle rushing. They took off and kept falling over logs — there's a lot of logs in that country — and of course they kept on taking fright at each other. You've got no idea what a mob of rushing cattle are like! One minute they're all lying down sleeping, chewing their cud, and before that first one that jumps can take two steps the whole lot are on their feet and galloping. They gallop in the opposite direction from where they got the fright, and it sounds like thunder.

> *Sunday.* Came to 1 mile past no 13 Bore & camped … Bullocks Rushed.
> *Monday.* Came to yellow water Hole … & camped   Bullocks rushed.
> *Tuesday.* Came on about 3 miles past No. 12 Bore & camped … Cattle Rushed   lost 45 head.
> *Wednesday.* Came on 1 mile past No 11 Bore … four boy[s] looking for cattle   picked up 20 head.
> *Thursday.* Camped at No 11 bore   Myself fishook & Wallaby went back looking for rest of cattle.
> *Friday.* Camped at No 11 Bore   Fishhook picked up 6 head of cattle.
>
> Humbert River Station Diary, entries for 14–19 May 1933

The Murranji was bad country for rushes. The timber and scrub in there made cattle nervous. If they got a fright they'd take off and if they took off, the dense scrub could stop you from getting a run at them too. I generally worked it that I went across the Murranji with a moon behind me. With a full moon, if they started galloping you could see where you were going when you chased them. Otherwise you'd be likely to gallop into a tree or an overhanging branch. That's what I was afraid of, an overhanging branch. The bullwady and lancewood doesn't bend, and if you hit a branch it'd knock you clean out of the saddle. When I took off after a beast I'd hold my arm in front of myself so I wouldn't be hit in the head, and I rode loose in the saddle. That way I'd only get it in the hand or arm and it'd be more likely to throw me back onto the haunches of the horse. I was very lucky. I hit a couple of branches but mostly I managed to evade them, more by good luck than good management I'd say.

Well every night they rushed we'd lose about twenty or thirty. We'd get out after them the next day and track them and pick them up again. The trouble was this held up our main mob. We were using water at the bores and not leaving enough time for it to be replenished for other mobs coming on our tracks. Usually big mobs of store steers were coming over from the Kimberleys and there was only about two to three days between them.

The cattle steadied down the last four or five nights before we got to Newcastle Waters. I thought, "They're right now," so I said to Roley, "I'll take the station plant back now." There were nine or ten of us in the camp you see. Roley kept going, but when he got to Alice Springs he delivered over 200 short! After that I thought, "Well, if they're going to bloody well lose them I'll take them in meself."

I took cattle to Queensland in 1935, '37, '39 and '41. Two mobs went in to Woodhouse Station near Townsville, one mob was sold to Vesteys and another mob was sold to some men by the name of McArdle and O'Sullivan, out from Stanford near Hughenden. The first lot I took in to Queensland was nine short. The next lot was eleven short, the third about five short and the last lot I think was about three short. A bit of a change from the mobs I'd had taken in by the other blokes! From the time I left on these trips I'd be kicking the pants of seven months before I got back to the station, so while I was gone I'd have someone on the station as caretaker to look after the house and garden — not branding or anything like that.

The first mob I took to Queensland was 1000 head of bullocks. All our gear was carried on packhorses and I'm not too sure if Roley didn't give me a start across the

Murranji as far as Newcastle Waters. When we reached the Nine Mile plain, nine miles from the Murranji waterhole, I sent the horse-tailer ahead with the horses — they were trotted on in, given a drink and brought back again. I gave the horse I was riding a drink from a canteen to keep him quiet.

It was about three o'clock in the afternoon and I was watching the cattle by myself when I saw some of them run forward and then back up, looking at something in the grass. I was about seventy or eighty yards away from them you see, and I thought, "What the hell is that there? I wonder if it's a damned snake, or a wallaby?" From the way they were backing away I had a feeling it wasn't a wallaby. A wallaby would jump and get away quickly.

I poked my horse in amongst them and the next minute I looked — it was about the biggest rock python I'd ever seen, one of those black headed snakes. They're pretty harmless, but this one was about twelve feet long. Usually you only get them about eight or nine feet. I didn't kill it. I let it go and it just poked its way through the cattle and around them, and away it went. And the cattle all had a look at him going past.

A bird of vast treeless grass plains ... At one time this pigeon was seen in countless numbers on the inland plains.

N. Cayley, 1984:247

Coming out of the Murranji, from about the Bucket waterhole right through Newcastle Waters there, as faaaar as the eye could see, there were flock pigeons. By God, they're good eating! They're all meat too, you know. I didn't have a shotgun so I had to shoot the buggers with a .22 rifle. I just sat on the side of the earthen tanks and shot them as they came in to water. The best feed I had was at Number 4 bore, near Eva Downs. I sat up on top of the bank with the .22, just before sundown. Gawd suffering Christ! I must have shot about twenty or twenty-five. I was running out of bullets!

From the Murranji we got out onto the Barkly, and we struck big rains at Brady's Grave between Number 3 and Number 4 bores. Four inches of rain fell there and we got no sleep that night. We were just riding around the cattle all the time while the rain was pelting down. On a count in the morning I'd lost about fourteen or fifteen head, but we got out after them and picked up all of them bar five. After lunch we pushed on to some high ridges on Eva Downs and then stopped the night.

He was undoubtedly one of Australia's greatest horsemen. He had energy, horse-sense and ability to cope with everything that turned up. When riding those that bucked he did not look the least like being thrown and I heard always the same from others that had seen more of his riding. "You never see daylight between Boomerang Jack and the saddle." His seat did not rise from the saddle, whatever happened.

H.M. Barker, 1966:124

That Brady's Grave I mentioned is the grave of "Boomerang Jack Brady", as he was known. Old Boomerang was a bloke who could *really* ride. At Wave Hill a horse came out of the yard there once, bucking and throwing its head around — look out when they start throwing their head like that because they're not looking where they're going you know — and the next minute it hit something, turned turtle and broke old Jack's leg. They turned round and pulled it here and pulled it there — I heard one rumour it was a compound fracture and somebody ripped a knife in his leg, opened it up and pushed the bone back. He told them to pack bags of sand all around it, and he was about two months or three months on his back and never shifted. Well the bone set, but ohh, when he walked he had a hell of a limp. Yet the old bugger still reckoned no one was better than him and he was still getting on colts. He'd yell — a savage old bugger he was — "Hold that bloody horse's head here, hold him tight! Grab him by the ear or something." And they'd say, "I don't think you should get on that horse Jack." He'd say to them, "I'm all right. I'm as good as ever I was," and he'd hit the bloody saddle and the bloody horse would fly away bucking. He never asked an Abo to ride a horse that he thought he couldn't ride himself, that was one principle he had.

> At about 2.30 am arrested John Brady Stockman at Wave Hill for attempted murder of Hunter Loder.
>
> Bow Hill Police Station Journal, 24 September 1918

Not long before I came into the country there was a gunfight at Wave Hill Station. Old Tom Simpson was there at the time and he told me about it. It appears there was a half-caste piece there, and like all women they'll always get someone to be chasing after them. Tom told me Boomerang Brady was one who was trailing her — I forget who the others were now. I suppose there was a bit of grog flying around as usual, but anyway, they got into holts and it developed into a war of rifles and pistols. Someone must have got shot because it went to court in Darwin. Shoot-ups happened quite a few times out there, but unless somebody got shot nothing was ever made of it. It was so much time and trouble to get blokes into court in Darwin that everyone just shut up and forgot about it — not like today where they can easily shift them into town.

Boomerang Brady had to go to Darwin and Tom Simpson was implicated too, because he had to give evidence.

The Judge asked him, "Mr Simpson, can you give me your version of events?"

Tom was a rough old bushman and he said, "How do yer mean, a 'version'?"

"Well what was your opinion of the half-caste girl that the row started over?"

Old Tom thought about it for a while and then said, "She's got a rattlin' fine figure, but she's about the ugliest thing yer ever clapped eyes on." Then he added: "She's like the peacock — the beauty's all in her tail."

I believe the court nearly turned inside-out laughing you know, and they couldn't stop! The Judge was banging the table. "If you people don't shut up I'll dismiss the court, I'll dismiss the court!" Well it was like asking those blokes whether they were going to go to heaven or hell — they just couldn't do neither! And the Judge said, "You people in the Territory, you're all queer!"

After the court case old Boomerang went back to Wave Hill. He had a half a dozen horses there and eventually he decided to hit out for Queensland. He reckoned he'd go home to see his sister. The police and everyone thought that was quite in order till someone asked him: "Say Jack, when'd you last see your sister?"

Jack replied, "Ohh, about forty-five or fifty years ago."

"You think she's still alive!"

In his bush ignorance he said, "Well, I don't know. She should be! I don't know what she'd have died from if she's dead!"

Several cases of beri beri, which I treated, resulted, I am sure, from personal neglect. The average bushman will not bother to boil a cabbage or vegetable when damper and cooked salt beef are available.

Dr Herbert Basedow, Adelaide Advertiser, 27 September 1922

Old Jack just managed to make it as far as Newcastle Waters. He was very sick and the policeman, Gordon Stott, looked after him for four or five days. Boomerang kept on saying: "I'll have to go, I want to get away Gordon, I want to get away," and Gordon kept saying, "You're a sick man." Jack's legs were swollen up with beri-beri from all accounts.

He pestered Gordon so much that Gordon said, "I'll tell you what I'll do. I'll lend you a tracker to go as far as Anthonys Lagoon and I'll write a letter to the policeman there. My tracker can come back from Anthonys and the policeman there will give you another tracker who'll take you on as far as the Ranken." When Jack got to the Ranken the second tracker was to come back and the next policeman would give him a tracker to take him through to Camooweal.

Anyway, old Boomerang Jack got out between Number

3 and Number 4 bores on Eva Downs Station, and he was very crook all the way there. Of course in those days there was nothing at Eva Downs. There's a scrub between the two bores and old Jack camped in there that night. In the morning the tracker didn't want to shift him, but it was a waterless camp so they got up on their horses and reckoned they'd go on to Number 3 bore. Old Jack only rode on about a mile before he said to his tracker, "By God I feel sick now," and the next minute he toppled off his horse and died.

The tracker stopped with him there — I suppose he panicked for a while, wondering what the hell to do. Then he pulled old Jack's swag off the packhorse and covered him up with it, and rode for three days back to Newcastle Waters. There he told Gordon Stott, "Him dead, him finish."

Gordon had to go all the way back to bury old Boomerang Jack. Jack's body was on his swag, so Gordon dug a grave right alongside and just rolled the swag and tipped him in. Then Gordon or somebody drove in some posts. Boomerang's sister, a Mrs Whittaker, sent a tombstone up marked, *In memory of Jack Brady* — if he'd lasted long enough he would have caught up with her. For years after, when the drovers were going that way they'd always camp at Brady's Grave and watch the cattle there. It was the only place for miles and miles where they could hang up the bridles and saddles at night.

Eva Downs used to be owned by Harry Bathern, or "Bullwady Bates", as he was known. When I knew him, old Bullwady had Beetaloo Station, and on my way back from Queensland after delivering my bullocks I used to go over and see him. I'd camp about eighteen to twenty miles off Newcastle Waters and in the morning I'd send the horses on in. I'd tell my boys, "You go on to Newcastle Waters and stop there tonight," and I'd cut across country to Beetaloo. I'd stop the night, and usually I carried a bottle of rum to give to the old fella. He'd have a good nip and by Christ he was funny. He had a mob of half-caste sons and daughters and a rough homestead there, and never ever got away from the place. He told me a bit about his history and how he got hold of Beetaloo.

Bullwady said he came out to the Territory when he was eighteen, and he hooked onto a lubra. He took up the Eva Downs block and got his water from a well, but it used to go dry on him. He had about 180 or 200 head of cattle and because of the bad water supply they had to shift the cattle off the place whenever there was a dry spell. They were taking them around Borroloola country and his lubra kept on saying to him, "Why don't we go longa Warungungo" — that's the present Beetaloo station — "There's proper big waterhole there can't go dry, 'im can't go dry."

Old Bullwady Bathern told me he decided to go and have a look for himself. His lubra was on a horse with their eldest son, Angus, who was only about fourteen or fifteen months old. He said, "She strapped a piller on the front of her saddle and put Angus on that, and we went across."

1907 ... Keeping south arrived at Bundarra Waterhole, where Harry Bates was camped with the Bulk of Eva Downs Cattle. We put the Cattle together & took them to Briggs Lagoon let them go & left.

W. Linklater, unsigned manuscript, n.d.

When he got there he was amazed. He said, "The shock I got when I saw it! It was about a quarter mile wide for a start and a mile long. I had a look at the water's edge and although it was on a black soil plain you'd never get one beast to bog there — the edge of the waterhole was real hard with rocks." He took out a lease, but when he moved over there Newcastle Waters decided to make out that it was their block and tried to hunt him out of it.

Those days the Northern Territory came under the South Australian Lands Department, so old Bullwady took off for Adelaide. He put it in the hands of a solicitor there and won it. As it turned out that particular waterhole was about sixteen miles off Newcastle Waters country. He said he couldn't get back to Eva Downs quick enough then. He mustered up all his cattle and walked them from there right through to Beetaloo, and he never looked back.

Poor old Bullwady. He was talking once about the bloody Abos. "Charlie," he said, "you're doing a big thing over there on Humbert on your own, but you want to be careful! You know, if something come over them they could spear yer, and the next minute they'd sit down along side of yer and cry over yer." I could understand that of them, you know. It's just their nature's way.

Anyway, to get back to my first droving trip to Queensland, from Eva Downs we went through to the other side of Number 2 bore. The rain was getting slacker as we got over that way. We went through to Anthonys Lagoon where I dipped the cattle, then on to Brunette Downs and Ranken River. At Soudan Station we turned off to the south and followed the Thompson River down, and went through to Lake Nash. That's where I saw about 20,000 crows.

Yeah! We were having dinner there on the big swamp on the other side of the homestead and I heard the bloody things coming. I listened, and I could see them way up, just little specks in the air. I said, "Them bloody things are travelling!" They circled round and round and at last down they came in a spiralling dive. God strewth — didn't they make a noise coming down! They held their wings back and the wind rushing through their feathers made a blurry sort of noise, a noise I'd never heard before except perhaps for a hawk diving. With thousands and thousands coming down at once they made a hell of a din. They landed on the edge of the waterhole and stopped all lunchtime, and provided amusement for my cook and the Aboriginals and everyone that was there. We started the cattle up and left about half-past two and we'd only gone about a mile when the crows decided to leave too. They were travelling east and away they went.

I saw another mob of crows like that at Humbert one year. Round about four o'clock one day there could have been anything up to 30,000 crows flew up in one big mob, definitely more than I'd seen at Lake Nash. I heard the blacks all singing out, so I went out and had a look. The flock appeared to be about half a mile long by a quarter-mile wide. They must have been a couple of thousand feet up and you could hear them quarking — everything was "quark, quark, quark". They circled the station and then decided to come down in a dive, just like those I'd seen at Lake Nash. They landed in about fifteen or twenty trees in front of the house and it took about three-quarters of an hour before the last one came to rest.

The noise of them was terrific! Just about sundown they took off again and I thought, "Oh good, they're off, they're going." They lifted off their perches in hundreds and flew round a bit — still quarking — and then they landed under a big sandstone range less than half a mile from the homestead, right on a big lagoon there. Well they did nothing but quork and quark all darn night, and nobody could get any sleep. All that had me worried was whether they were going to shift along in the morning. I didn't want damned crows there with all that noise. Just as day was breaking I walked outside and they started to take off. The leaders came down to the homestead, about 100 feet up, and the noise was deafening. They kept on circling there and they seemed to be waiting for the tailers to catch up. They circled the station six or eight or a dozen times, climbing higher and higher and higher — it took them the best part of an hour to get right up. Then all of a sudden one hit out to the east. About fifteen or twenty followed in a "V" formation and then the whole lot took off.[1]

This first mob I took to Queensland had a lot of old pikers in it. Old bullocks are very heavy which makes travelling hard on their feet, especially on stony ground — a young bullock can travel a lot easier or better than an old fella. Quite a few of the old bullocks I had went lame and I ended up shoeing nineteen before I got them to their destination. Actually, when you shoe a bullock it's called cueing them, not shoeing. Bullock shoes are horseshoes cut in half — you can use horseshoes that are worn out. One half is nailed to each side of the bullocks' cloven hoof and two nails are all that's needed to hold each one on.

It's an amazing thing, those bullocks, you'd see that one was very lame and decide to shoe him. You never caught him around the neck with a rope because he's got all the power under the sun and he'd pull you all over the place. I only did it to one or two and it suddenly dawned on me to catch their hind leg as they walked along. When they're lame they're dragging their foot all the time so they're all back on the tail of the mob. I had two bronco mules and bronco gear with me and I'd catch the lame bullock. They get that quiet you can ride along and put your hand on their rump you know, and I'd just catch him by the hind foot and stretch him. If he went to fly round I'd stretch his leg out and he'd more or less stand. In the meantime somebody else would come round to the front and catch him by the front leg, stretch him, and over he'd go. Somebody else would hop in then and grab him, and hold him down. It's just as easy to do it right out on a plain as anywhere else.

I wouldn't shoe all four feet. It was always a hind leg that went lame and needed shoeing, so I'd do the lame foot and then turn him straight over onto his other side and do the other hind foot too, because if one foot was very tender he'd be saving it, and putting his weight on the good foot. If I didn't shoe it as well, the good foot would go lame. Well for three or four days afterwards those bullocks would still be on the tail, and then all of a sudden you'd notice they were gone. Your first reaction was, "Gee, I've lost those bullocks." You'd start looking for them and you'd usually find them up in the middle of the mob. Not one or two, but *every* bullock that I shod was the same way, and without fail, in about a week's time those same bullocks would turn up within the first six or eight or ten that were in the lead. So it just shows how it knocks them back in their walking when they're footsore.

The next time I took bullocks in, the head stockman on Woodhouse told me: "Those Territory bullocks, you know, we're still gettin' them here with shoes on." You can shoe a horse and the shoe might stop on six weeks, but he'll usually have lost it by then. Some of those bullock shoes must have been on for the best part of eighteen months! If anyone had told me beforehand that they'd stay on that long, I wouldn't have damn-well believed them.

On my third trip across the Murranji I had to do a fifty or sixty mile dry stage. One of the bores was out of order, and the Murranji was only half-full. Years afterwards they put a bore down at the Murranji waterhole. "John Government" did it, and it took them long enough, but it was very handy there at the last. Before that, if the Murranji waterhole was dry you had to go from Number 11 bore right through to Number 10 without a drink for your bullocks, about fifty miles.

To get my cattle across that dry stage I gave them a good drink round about mid-afternoon, and then started them up and walked them till about dark. The wagonette was waiting along the road and when we got to it we had tea. Then we walked the cattle till about ten o'clock and pulled up and camped. The next day was a dry day. We kicked off about daybreak and kept them going till about two o'clock. It was pretty hot then and they were wanting a drink, but they couldn't get one till next day.

Our horses needed water so I either had to send them back to the last water or, if there was enough water on ahead, I sent them that way. After they were watered they were brought back and met us on the road. The next day the cattle got in to water round about two o'clock, and by gees, we didn't have to ask them to have a drink either! If I could work it, I'd only go out about another half-mile from the bore to make camp. You were only supposed to give your cattle one drink at each bore, but if there were no other cattle pushing up from behind I'd whip my cattle back again in the morning and give them a second drink. Then away I'd go to the next bore.

I was always pretty wary of those tanks because the water would be in them for so long, and you often got a lot of dead animals in them. Different blokes would throw big branches in so that birds could land and get a drink, but sometimes I've seen so many dead birds in a tank you could hardly put your billycan in. I always carried extra water in a canteen that I had, just for drinking and for our meals.

Coming back from Queensland once I pulled in at Number 12 bore to camp the night. I had a look in the tanks there, and I looked again, and I said, "Jesus! There's a couple of damned dingoes drowned here!" I could see then how they got caught. They'd walked out about eight or nine feet along a six-inch overflow pipe, but when they got to the edge of the tank the water was down about two or three feet. I suppose they bent right over and when they couldn't get a drink they eventually either jumped in or fell in. Of course, once they were in they damn-well stopped there. They couldn't touch the bottom, and they had no way of getting out you see. They were still pretty high of course, but they were floating and you could still make out they were dingoes.

Later I struck Jungari Crouch at Top Springs — he was just knocking around with three or four bagman horses. I told him about the dingoes and he said, "Yeah, I come along there the other day at Number 11 and I seen a ripple in the water, and there was a dingo there. It was just about drowned." Jungari told me the dingo was just feebly swimming around, so he waited until it came round his way and then grabbed it by the neck. He said it never even snarled at him, so he pulled it out and hit it on the ground and killed it. And he said, "I got thirty bob for his scalp, the easiest scalp I ever caught!"

People used to waste water at those bores. I forget how many thousand cattle used to come across the Murranji in a year, but once Wave Hill started moving cattle they'd send in anything up to 10,000 or 12,000, and there were other places too. My little mob of 1000 head was only a drop in the ocean in comparison with some of the other places. But some of the drovers could be very careless and wasteful. They just thought there was an endless supply of water. Well it may be endless, but it only comes out of the ground in a certain quantity.

The drovers would send in their cattle in mobs of about 200 — they wouldn't want the whole mob to rush the troughs at once because they'd be ploughing over each other and falling into the troughs, and goodness knows what. They'd take off the bloody floats to get a full six-inch flow rushing into the trough. But one lot of cattle might be a bit slow coming in because they were so far back, and of course if no one was there to watch out and shut off the flow the troughs would start overflowing onto the ground.[2]

Most of the bores had 30,000 gallon tanks made of galvanised iron, and the drovers used to write messages on the sides. Someone came along with a bit of paint and wrote a poem on the side of one tank. I think it was done by an accountant who used to work in Max Schobber's store at Newcastle Waters. The poem was to do with the wastage of water by the drovers, and the death of two pilots, Hitchcock and Anderson, in the desert to the south. I thought it was very good. They were Metters tanks with the brand-name on them, and the poem was written around the name on the tank like this:

> For this
> METTERS PATENT STEEL SQUATTERS TANK,
> You have the North Australian Commission to thank.
> Someone else's job to fill it,
> Yours the privilege to spill it.

> The Kookaburra stands without a cover
> Fifty miles from where this trough flows over
> Of the water you've wasted just a gallon or so,
> Would have saved the lives of Anderson, Hitchcock and co.³

After the war I was coming along there, and around the other side of the same tank someone had written in big scrawling letters: "Wanted by Hitler: Schobber, Schultz, Althaus and Zigenbine." They were all local identities at the time. By God, I had a good laugh over that you know. Anyway, I noticed that each of those blokes listed was boss of his own set-up; they weren't just ordinary blokes — they must have had a certain amount of brains to get where they were!

When we got in to Lake Nash we discovered that the road ahead wasn't the best. There was no feed down the river and the stock inspector's advice was to go through to Mount Isa and follow the West Leichhardt River down to Dajarra. At Dajarra I got the bullocks onto the train for Woodhouse station and someone took my horses back to Lorraine Station. My sister Phyllis and my mother were there to meet me at Dajarra, and another young woman called Dolly Creery. My mother had just about raised Dolly, and she was good friends with Phyllis. After I delivered my bullocks I went through to Ayr with them.

Whilst breaking in his horses at Woodhouse Station he attended the Townsville rodeo and represented the N.T. He was sixth in the buckjump contest for the Australian championship and won two smaller events. Between 40 and 50 riders contested the Australian Championship.

The Northern Standard, 24 July 1934

While I was over at Ayr I ended up in a couple of rodeos. I got thrown in Ayr, but I rode everything in the St Johns Rodeo in Townsville. Forty-odd competed in the camp draft and I got second place. Throwing and tying the bullocks there, they gave me a six-year-old bloody bullock and no one to come up on the offside of me. I got a hell of a buster out of that. I left my horse and grabbed the bull by the horns and he threw me to bloody hell. I hung to him for about thirty yards, and at last I let go, and he threw his head up and sent me clean in the bloody air.

> *An airmail letter to Mr W. Miller, from Charley Schultze, manager of the Humbert Station ... advises that he is starting out from Woodhouse Station, close to Townsville, with 150 head of horses, for the Humbert Station, a long trip of about 1700 miles.*
>
> The Northern Standard, 24 July 1934

I wasn't getting any wages for the work I was doing on the station at that time, but Cunningham paid me for droving, so after I delivered my cattle in Queensland I'd put that money into horses. I'd buy about thirty or forty, fifty sometimes, and I'd break them in and take them back to Victoria River. Together with my plant horses I'd go back with about 140 head. I'd only pay about £2 a head, and I'd get a fiver for them — I'd clear £2 a head. VRD took one lot, Newry another, and another I sold to a drover.

9 A Lot To Be Thankful For

Charlie Schultz was one of them good man on the horses and cattle and the mustering. He was a smart bloke. He's a really bushman, I think he's a half a Ned Kelly or something.

Doug Campbell, quoted in D. Rose, 1991:135

... I have a lot to be thankful for, as regards these abos, and realize without their aid, through lean and hard times — and without any complaint — so to speak, I would not have been able to carry on without them and their help.

Charlie Schultz, 20 September 1948

The Australian natives were on the move all the time. They might camp for two or three days and when they had everything killed in the vicinity of four or five miles around, they'd take off again and go to the other hunting grounds. When the wet season came they just had their bark gunyas. Their own commonsense told them they didn't want a permanent base and according to my reckoning that's why they never made a house. It was very hard to make them sit down permanently. They were people that were on the move all the time to get their tucker, and that's why the men made such good stockboys. Being such good hunters, they knew where to hunt up the cattle.

In 1935, long before it became a government regulation to build accommodation for your blacks, I put up three round grass huts for them. I thought the design out in my head if I remember rightly. They were a good ten or twelve feet across with wooden frames and covered with thatch. A good timber to put in a roof is peartree. If I could have got it, the frame would have been leichhardt pine, but to get leichhardt all the one size is fairly hard to do. I probably

A family shot taken at Woodhouse station in about 1914. Charlie is the little boy scratching his nose. To Charlie's left are his father, sister and mother. *(Photo: Schultz collection)*

Four of the Schultz brothers. *From left:* Albert, Dan, Charles (Charlie's father) and Billy. Missing are Henry and George. *(Photo courtesy Darky Pollard, Ravenswood)*

Charlie and his father at Woodhouse station in about 1925. *(Photo: Schultz collection)*

The homestead and storeroom at Humbert River a few years before Charlie arrived. The boab tree, planted in Billy Butlers' time, is still flourishing. *(Photo: Feast Collection)*

Albert Schultz, Jack Roden, Jack "Broken-leg" Farrel and Billy Schultz at Humbert River station in 1922. Shortly after this photograph was taken Roden became the bookkeeper and store-keeper on Victoria River Downs. He was there for twenty-six years and had a long association with Charlie Schultz. *(Photo: Feast collection)*

"The first seven years were the hardest." Young Charlie in about 1930. *(Photo: Schultz collection)*

The fencing camp in Riley Pocket, September 1928. Charlie created a bullock paddock by running two strands of wire across the mouth of a cliff-lined valley — the first fence on the station. *(Photo: Schultz collection)*

Inside Humbert River homestead in 1938, when it was still a bachelor establishment with corrugated iron walls and an antbed floor. On the left is Dave Fogarty reading a *Walkabout* magazine. Charlie is sitting beneath a gun rack, typing a letter. On the right is Dave Magoffin, in his R.M. Williams boots. *(Photo: Magoffin collection)*

In 1910 rumours reached the Timber Creek police that Brigalow Bill had been murdered. This photograph shows Mounted Constable Holland at the blood-stained door of Brigalow Bill's hut. *(Photo courtesy John Gordon)*

Mounted Constable Holland and tracker with four of the alleged murderers of Brigalow Bill, at the end of a 100-mile walk to Timber Creek. *(Photo courtesy I.J. Raymond)*

The "Big House" at Victoria River Downs in 1925. This homestead and an identical one at Pigeon Hole outstation were built early this century, from timber cut on the station. *(Photo: C.E. Schultz collection)*

The treacle and bun race was a "spectator sport" organised at the VRD Christmas sports in the 1930s. *(Photo: Buchanan collection)*

Charlie Schultz, Dave Fogarty and Roley Bowery arriving at VRD for the Christmas races in 1936. *(Photo: Walker collection)*

The Wimmera Nursing home at VRD in about 1930. The small roofed building at the left was an annex for Aboriginal patients. *(Photo: Roden collection)*

Waiting for the outcome of a protest by Frank Spencer against Charlie's horse at the VRD races around 1935. Alf Martin is on the left, dressed in white clothes. The first three horsemen from the left are Frank Spencer, Roley Bowery and Bob Nelson. *(Photo: Roder collection)*

At the VRD races in 1950. Frank Spencer, Hessie Schultz holding the VRD cup, Hartley Magnussen and Charlie Schultz. *(Photo: Mettam collection.)*

"All they had were a couple of paperbark sheds." The homestead at Gordon Creek outstation in about 1930. *(Photo: Roden collection)*

The VRD mob. Standing are Charlie Arnold, Ted Martin, Dan McAlvary, Jack Knox, Roley Bowery, Snowy Shaw, Bob Nelson, Frank Reynolds and Frank Spencer. Sitting at left is Jim Martin. *(Photo: Buchanan collection)*

"I'd always attend the musters with Frank Spencer." Frank Spencer and Charlie Schultz at a temporary stock camp near Mount Warburton. *(Photo: Roden collection)*

Charlie Schultz and Frank Spencer during a joint muster on the Victoria River Downs–Humbert boundary near Mount Warburton. *(Photo: Roden collection)*

Humbert River blacks. The tall man in the dark shirt is Humbert Tommy who fought with the police tracker in 1937. To his left is Humbert Jack. Right in front at the far left is Snowy and next to him is Daly, two of Charlie's best stockboys. *(Photo: Schultz collection)*

Aboriginal women at Humbert River. *Left to right:* Biddy, Kitty, Violet, Fanny, Molly, Bunjy. *(Photo: Roden collection)*

David Magoffin, Charles Schultz, Dave Fogarty, c.1938. *(Photo: Magoffin collection)*

The Humbert River mustering plant about to leave the homestead, around 1935. *(Photo: Schultz collection)*

A thousand head of Humbert River cattle on their way to Queensland in 1935. Eighty miles from Humbert they are still on VRD, and about to start on the notorious Murranji track. *(Photo: Schultz collection)*

Charlie's stockboys in 1935. In the background are the huts Charlie made, one completed and two yet to be covered with grass thatch. *(Photo: Schultz collection)*

Bringing in the first bush cows to establish a milking herd in about 1936. In the background at the left are three thatched huts made by Charlie. The stone building in the centre is a store and at the right is the homestead. *(Photo: Schultz collection)*

One of Charlie's stockboys, Fishhook, outside his grass-roofed hut, in the late 1930s. (Photo: Schultz collection)

A wagon load of flagstones outside the station store in 1939. The man leaning over the wheel is Mahnikee. Opposite him, at the back, is young Snowy. Seated on the wagon is the VRD bookkeeper's daughter, Doris Roden, believed to have been the first white woman to visit Humbert River. (Photo: Roden collection)

Humbert River homestead as Hessie saw it when she came to live there as Charlie's wife in 1941. The stone building at the right is the bathroom. *(Photo: Schultz collection)*

A bush birthday party: Donna and Betty in front; Charlie, Hessie, Roley Bowery, and George Temble behind. *(Photo: Schultz collection)*

The family portrait: Charlie, Hessie, Donna and Betty in about 1953. *(Photo: Schultz collection)*

Hessie Schultz watering the garden which supplied the family and the Aboriginal workers with fresh vegetables in the late 1950s and 1960s. The building is the station store. *(Photo: Schultz collection)*

Donna, Larry Johns and Betty, doing school work at Humbert in about 1960. *(Photo: Schultz collection)*

ert Drew's donkey team outside the Timber Creek police station. From the beginning of ettlement to the late 1930s, almost all the supplies for the stations were brought to the epot by boat and then shifted inland on wagons such as this. *(Photo: Mahoney collection)*

oaded camels outside the VRD store. Camels were used on VRD for many years to take upplies to Mount Sanford outstation, and elsewhere. Several other stations in the district sed camels and some even sent camel teams to the Depot to pick up stores. *(Photo: Feast llection)*

Pushing Charlie's first truck across the causeway at Newcastle Waters, 1933. Roley Bowrey is wearing the hat. *(Photo: Schultz collection)*

Charlie and Hessie meeting his mother on her arrival on the first plane to land at Humbert River, in 1958. *(Photo: Schultz collection)*

"He was a bastard of a man." Wason Byers striking a casual pose in about 1940. *(Photo: Schultz collection)*

Wason Byers (left) and Jim Martin counting the cattle before Byers started them on the road to Queensland. The photo was taken across the river from Dashwood yard in about 1940. *(Photo: Schultz collection)*

Stud bulls being unloaded at the VRD airstrip in June 1951. These were the first cattle to be airlifted into the Northern Territory. *(Photo: Schultz collection)*

atering a mob at the Dinner Camp Hole, above Humbert River homestead. *(Photo: Schultz ?ection)*

tering cattle in Ivnors Pocket. Dave Fogarty is the horseman. *(Photo: Schultz collection)*

Charlie Schultz looks out at the ranges near Ivnors Pocket, on Humbert River. *(Phot courtesy Marie Mahood)*

"One of the best horsemen in the Ter tory." Charlie Schultz in the 196 *(Photo: Schultz collection)*

used a lot of bloodwood. I found that it takes a hell of a lot of grass to thatch a hut. I took the boys and gins down to the river and we cut grass about three to four feet long with butchers knives and a grass-hook. When I covered the frames with grass I started from the bottom. I cut a greenhide strip about three-quarters of an inch wide and ran it through an arsenic solution. If I didn't treat the strips, beetles would eat them, and eventually they'd snap and the grass would fall out.

I'd pick up a bundle about two inches in diameter and get the greenhide and take a turn around the grass. Then I'd go on to the next one. Each bundle spread out a bit on the end to throw the water off, and there was a fair pitch on it so that the water would run straight down. As time went on the greenhide tightened up and really held. I made the roof on a few of the huts extend right to the ground and a few had a gap of a couple of feet. Some of the Abos wanted them right to the ground and others said, "Oh, too hot." I made three in the first year and three the next, one for each family. What had me jiggered was how I was going to cover the hole at the top. You do this sort of thing by trial and error. I made a wooden block or plug of leichhardt pine or fig tree, about a foot by six inches. Both those timbers are lightweight. I put two holes through them at right-angles to each other and put wire through to fasten them onto the battens of the timber frame. The last layer of grass bundles was tied to this block with light tie-wire.

Before the posts went in the ground I treated them with arsenic and caustic soda. To do this I bored a slanting hole with an auger bit about six inches deep in each post. I filled that up, put a wooden plug in, and sawed it off so the piccaninnies couldn't pull the darn thing out you see.

I was always rubbing it into them about the fire danger. I said, "Look out for fire now, it's the piccaninnies you want to watch. If it ever catches fire you're gone." I think they could've scrambled out once they saw it starting to go up, but it would have been every man for himself. I can just imagine it — gins would be flying out, dogs'd be barking — the dogs always have to have a finger in the pie. Funny how dogs grow up in camp like that on a station; if anything goes wrong they can scent it straight away. If anyone has a fight or something's not natural, they wake up straight away.

> 1/2 lb borax (if possible)
> 3 lb caustic soda
> 15 lb fat
> 2 lb rosin
> cup full of Kerosene
> a few grains of condys
> a kerosene tin of water
> Mix up the lot & boil till shingy "toffeyed" stirring all the time.
>
> Humbert River Station Diary, 1933

In the early days at Humbert it was difficult to get things onto the station, and we used to do without a lot of things or else we'd make our own. One thing we made was soap. The ingredients were fat, caustic soda, rosin and kerosene. After I got married my wife used to have anything up to about four or five hundredweight of soap and the longer you kept it the better — it dried out and had more kick in it. If it's not real hard it doesn't last so long. By gee, you used to get a good lather with it, but the caustic soda was a bit hard on your clothes.

Every Saturday the blacks would come up to where we'd give out the rations and we had the soap there in boxes. If one boy had a couple of lubras you gave him about three pieces of soap. By gees they were heavy on soap. It was a good thing because they used to keep themselves clean, although there was a certain amount of waste with them too.

On most stations it was the cook who made the soap. If you got a good patient cook, every time a beast was killed the first thing he said was, "Bring that fat down, I want it for soap." A good cook would always use all available fat, but some of the buggers, after they got a certain amount they wouldn't bother. They'd have that "cook feeling" — they'd be too tired to do it. Of course in later years we could just get anything from town.

> *Accommodation: Blacks. Galvanised iron shelters. Paper bark humpies. Much improved arrangements for natives than those usually obtaining on stations. Every indication of thought and care for the welfare of natives.*
>
> Ted Evans, 4 August 1950

When I got onto my feet financially I made tin huts for my blacks. I cemented the floors and put big windows in them, and I made the bottom of the windows level with the bunks, so that they'd get any breeze, and you could prop them open with a stick or you could keep them shut. They all had cyclone bunks that I got from Katherine after the war. There were big cyclone bunks everywhere then and every time I went into Katherine I could always manage to bring back a half a dozen. You'd pick them up in the rubbish heaps. I put a door on the huts too so that when they went away they could lock them up. Otherwise they were interfered with.

I found the blacks very good with the huts and they looked after their clothes better too. Previously, when they were going away on walkabout, they'd hang their clothes up or leave them in a bark humpy where the white ants'd be eating them.

I got a contractor from Mount Isa to put up a great big kitchen with a stove and a big long table and seats. I had lubras there to cook the tucker for everyone, but my wife had to watch them. They wouldn't do anything for themselves — you had to stand over them. They cooked mostly stews and curries. We always had plenty of vegetables there. I'd give them an enamel plate at the beginning of the year, and a pannikin, and if they asked for a second plate or cup they never got one, but they were just as happy to drink out of a damn jam tin, you know. Out in the mustering camp they usually had jam tins, the mongrel cows! You'd ask: "Where's your pannikin?"

"Oh I bin lose 'em," or "I bin leave 'em there, 'im bin stop at station. Plenty *sham* tin around here," they'd say.

So they'd have jam tins for a pannikin, and when they were finished with it they didn't even have to wash it. They just kicked it into the grass.

Later on I put water taps and one electric light on a pole in the middle of the camp. Water and electric light was laid on to the kitchen too. Then right against the house were showers with cement floors, one for the boys and one for the lubras, and on the other side of the turnout there were septic tank toilet systems.[1] After I left Humbert, someone pulled the bloody things down.

> *A matter the writer has been going to write to you about now for some considerable time, concerns wages for my Stock Boys.*
>
> *The position is, for about the past two years — as soon as I started to get on my feet, and clear this place of the Banks overdraft, I have been paying each 5/- a week pocket money, and now that we will be sooner or later paying them wages, I would like to point out that arrangements of some kind, may be made, for them to receive some of this, instead of it all going into a trust fund in Darwin.*
>
> Charlie Schultz, 20th September 1948

As well as putting up huts, I decided to pay my boys and lubras too. At that time they weren't getting paid on VRD, but I didn't care a bugger what the VRD whites or anybody thought about me paying my blokes. I just paid them and that was it. On VRD the old Abo knew that he wasn't going to get money and he just carried on under that assumption.

Later on the government brought in regulations about paying Aboriginal workers and I used to pay them by cheque, just in case the Aboriginal Department came back to me and said that I didn't pay them this, or didn't pay them that. This sort of thing happened on occasions when there was money owing to the blacks. A patrol officer came along one time and went through the books, and he bamboozled me with science on figures. He reckoned one of my boys, Toby, had some money coming to him. "He's got thirty-six quid coming to him," he said.

I said, "Okay then. I'll give it to him now and you can be a witness."

"Oh no, give me that. I'll take it down to Darwin."

"What the heck is the good of it going down to Darwin? He's bred and born here." He had a lubra then too — old Daisy. I said, "I'll give it to him or if he likes he can leave it with my wife and she'll give it to him whenever he wants."

"Oh no," he said. "No. I have to take it down to Darwin."

I said, "Once it gets down to Darwin they might never see it again."

"Oh, we'll keep it in trust."

Toby never saw that again, and this is the very same department that was always on the pastoralist's back about not looking after the Abos. I can assure you, when you clothed them and their dependents in the camp, and kept them in tucker and tobacco, you were up for a few bob.

We'd pay them monthly and we'd give them a cheque so it could be traced. On average, each one had about twenty or twenty-five quid coming to them every month. Well, before the mail plane came out they'd give the cheques to Hessie and tell her what they'd like ordered from Cox's Store in Katherine, but with the orders they put in they clearly thought their cheques were worth about £1000! Oh God, they used to drive her off her head with what they'd want! Each one would end up with a list of about ten or fifteen items and Hessie would send this in to Cox's store. Each order was put in a carton with that person's name on it, care of Humbert River station, and when they arrived back on the plane we'd have a truck ready, and all the cartons were taken down to the office. Of course, my wife used to let them open their own parcels and they got a great kick out of it. She'd look at each carton and call out, "Nina?"

"Here."

"Daisy? Lookout, this one belong Daisy. Jessie?"

She just handed them out and, Jesus, they were like kids with a new toy!

> ... I would like to record that, as always, there is an air of progressiveness and contentment on Humbert River that extends to the native camp and I feel certain that Mr and Mrs Schultz will continue to be leaders in showing appreciation of the work of their native employees and in providing facilities for their welfare. As a Native Affairs official, it is a pleasure to visit Humbert River Station.
>
> Ted Evans, 4 August 1950

I will say this about VRD though. In spite of what some do-gooders claimed, they couldn't have treated or fed their Aboriginals or the white stockmen better, especially in Alf Martin's time. It didn't matter what they wanted, all the cook or head stockman had to do was send a request to VRD and out it came. In fact there was a hell of a lot of waste going on.[2] The rations came every six weeks. The cook'd think: "Oh we're going to have an oversupply," so the first blacks who came along, he'd give them what they could carry away. You'd go into the blacks' camps and they'd have bottles of tomato sauce there, and tins of jam — it was ridiculous the clothes and rations they were given. I've seen anything up to five jars of chutney just left behind in the stock camp. If there was no way to carry it without the jars breaking, they drove the dray away and just left it there. At different times when it was convenient to carry I'd take some leftovers home. Times were tough there on Humbert — at times I didn't know where the next bag of flour was coming from. Old Tom Simpson at Gordon Creek turned around once or twice and said, "I've got a spare bag of flour, Charlie. We've got more than what we can eat here. I could send it down to the blacks but if you go down there you'll see they've got dampers not touched under the trees."

Jesus the buggers were looked after! The police raided one of the VRD camps there one time looking for somebody and when they went through the swags they found a sugar bag full of blacks' tobacco. Of course, anyone that came in from out bush, blacks or anyone else, they just handed tobacco out to them.

The station got their horseshoes by the ton and every six weeks they'd send out two hundredweight to each stock camp. The camel team would come along and any stuff that was likely to get knocked around in the weather they'd put under cover, but the horseshoes — "Oh, sling them over there against that tree." You'd go to a stock camp and you'd see a heap of horseshoes under a tree with the bag rotted away — been there for donkeys' ages and just not touched. If you went on to the next camp there were just as many shoes there too. I got a few heaps — but that Frank Reynolds, in spite he'd gather them up and push them in to the station, or leave them at Gordon Creek. Still, I don't think I bought a horsehoe the whole time I was on Humbert.

I used to take my wife for a drive in the Land Rover every afternoon, just to break the monotony for her. We went out one afternoon and as soon as we left, a big fight started in the blacks' camp. There often used to be fights when Hessie and I were both away from the homestead.

By the time we got home the sun was nearly down and as we crossed Peter Creek we saw all the blacks standing near the homestead. Hessie said to me, "What the heck's gone wrong now?" And as a joke I said to her, "I hope to God nobody's dropped dead!"

As soon as we pulled up at the garage they walked over and one of my best boys came up, a boy called Daly. He was very agitated. He said: "*Sharlie*, you can kick'em my arse." My first reaction was that he'd gone mad. I said, "What's the matter?"

He said, "That one Barney, we bin twofella have'im row and I bin hit'im longa stick. You can kick'em my arse *Sharlie*, you can kick'em my arse."

"Just a minute! What the hell's happened to Barney?"

"I think might be him dead. I bin hit'im longa neck with a stick."

Long Barney was a boy that didn't start stockwork until he aged, and he was still a bit of a myall blackfella. He was a good worker while you had him, but you never knew when he'd go bush. He'd just leave all his clothes behind and take off. All he'd grab was a loin rag, two or three boomerangs and a couple of spears.

Anyway, Daly said Barney had gone over and thrown something at Snowy, or hit him with a boomerang, or some bloody thing. Snowy was Daly's brother, so Daly belted Barney. The part that always rocked me was that up until then, Daly and Barney were damned good mates. Then Daly hit him across the back of the neck and down Barney went, and by Jesus there was no mistake — it killed him instantly.

I said to the wife, "Go down and have a look at Barney while I get things straightened up here." I followed along a bit later and I met Hessie coming back. She said, "Barney's dead all right. You needn't go down. You should get the police on the radio." I rang up John Gordon about it and told him what had happened.

Gordon said, "There'll be a court case over it. Do you think you should chain him up?"

"No. I told Daly not to be frightened and run away because that'd only make trouble for him. When you live with the Aboriginals as long as I have you know when you can trust them."

So he said: "Right. The onus is on you to keep him there. I'll get down there as soon as I can. I'll have to arrest Daly and we'll go straight in to Katherine."

We got Tiger Lyons to defend Daly, and there was a

Native Affairs officer there too. I told the officer all about it and he thought he'd play a big role in defending Daly, but old Tiger Lyons ignored him and asked me. I gave over the full issue and they dismissed the case.

Outside, I said to the officer, "I'll tell Daly now, because he's living in suspense." I went over and said, "You're all right Daly."

"I'm all right *Sharlie*?"

"Yeah. Your case was dismissed. You're right."

"I can go back to Humbert with you *Sharlie*?"

"Yeah, you can come back."

Daly wanted to go straight back to Humbert with me, but this Native Affairs bloke wanted to take him down to Darwin and keep him there. He reckoned there'd be payback and all this sort of bloody bullshit. I said, "Look. You're taking one of my best boys away. All I have to do is go out there and line 'em up and say, 'Listen. None of this pay-back business or you'll be in bloody trouble. There's been enough trouble here, so shut up and no more bloody rowing in the camp.' " That's just what I did, and there was never the slightest bit of trouble after that.

A pleasing feature of the food supply at Humbert River is the amount of fresh milk made available to all natives. There are several milking cows on the station and these are milked daily by three of the female employees. After the household requirements are taken out the bulk of the milk goes to the native camp.

Ted Evans, 4 August 1950

I always thought there's nothing like seeing a station with quiet cattle around the house — a lot reckon they're a damn nuisance, but give me the old quiet beast every time. Before I was married I started a milking herd there on Humbert. When I'd go out on a mustering run and see a cow with a heifer calf, I'd cut them out and bring them back to the station. The first lot I got over at Ivnors Pocket. We kept the calves yarded up in a pen and for a start we had to lock the cows up too, but once they realised their calves were there they settled down. Then I let the cows out to graze during the day and put one or two gins on horses to tail them. Later one or two boys went out on foot to get them used to being driven up by people. They took fright at first and I had someone on a horse to block them.

Eventually I had a cow shed with three bails and a cement floor, and I generally had three gins milking about twenty-six or twenty-seven cows. Being only bush cows they didn't give anywhere near the milk that southern cows give. I'd say from some we'd only get about three or four cups-full at the outside. As time went on I improved the milkers by getting in an Illawarra shorthorn bull and three Illawarra shorthorn cows from Queensland.

I had a milk separator too, and I used to shame Victoria River Downs. We'd go over every fortnight or so for our mail, and I'd take over two big three-gallon buckets — one full of cream and the other full of butter. The first lot I took over, Jack Roden said, "What the hell are you doing with this?" I said, "We've got too much on hand so I brought it over to show you blokes how to make butter." They were always struggling over there, they could never make a thing. They'd only have about three cows in for a start and then they wondered why they weren't getting any butter. And the butter they did have melted into an oil.

At first I kept my butter cool in a Coolgardie safe, but it wasn't too long after that that I got a kerosene fridge. I'd separate the milk and get about a gallon bucket full of cream. The blacks took all the rest of the milk away. They used to hang around there with a square four-gallon kerosene tin. Jesus, they thrived on it!

Well, that's how I first got my quiet cattle going. At the end I had about 600 head just running around the house there, and up Muldoon Creek — they were everywhere. Then of course, we always had the goats for milk through the wet weather. When I sold out, the new owners didn't bother about the goats and they just went bush where I suppose the dingoes cleaned them up.

10 Wife and Kids

> *Mr & Mrs C N S arrives at humbert River Station for dinner. All the blacks come up to have a look at the new 'Missus'.*
>
> Humbert River Station Diary, 18 July, 1941

My sister Phyllis and Hessie Graham were great friends. Hessie came from a farm near Ayr you see. When I was taking bullocks into Queensland, my mother and sister always motored right out through Cloncurry and spent the last ten days travelling with me, droving. They usually brought somebody with them and Hessie was keyed up to come out this particular time, but a few days before they were to leave Ayr her father took very bad and inside of ten days or a fortnight he died. Well, the funeral and the Will were coming up, so she had to call it off. Another friend of my sister's, a girl called Dolly Creery, took Hessie's place.

That year Cunningham took the bullocks off me. He said, "You bring them in here and I'll guarantee you a buyer." I had to walk them 1000 miles to either Dajarra or Kajabbi, the head of the railway there, and then truck them down to Woodhouse. My mother said, "Come on, you'd best come on in and see your dad. You can come back again then." I left my horses in Camooweal and went in and saw my father, and that's when I met Hessie.

Phyllis was all the time talking about Hessie, and the second night I was there I took Hessie to the pictures. About twelve months later I was in at Woodhouse once more and I caught up with her again — she was very keen. After I came back to Humbert my mother and Dolly Creery came out for a holiday. Later that year Hessie was going across to Singapore and she detoured out to Humbert for a couple of weeks. In those days if you were flying to Darwin you had to go down to Brisbane, Sydney, round to Melbourne, Adelaide, then north through Alice Springs. Well, she turned up at VRD on the little mail plane that came from Daly Waters across to Wave Hill, VRD and on to Derby. Something happened and I couldn't get over there in time to meet her, but Mrs Martin looked after her all that day and the next day. She was very good, a lovely lady Mrs Martin was, you know. We had a car at Humbert, but the Wickham River was still running just then and I didn't like asking VRD if they would run Hessie out from VRD to the Humbert junction. I knew that she could ride so I took spare horses over.

I said to Hessie, "How long since you've been on a horse?"

"About six or eight years, but I can ride."

I thought, "Oh, that's bloody lovely."

We went first from VRD to Frank Reynolds' place at Gordon Creek. That was twelve miles. Frank was out mustering, but Mrs Reynolds was there and we stopped the night. I was hating to think what the next day was going to be like because Hessie was pretty tired when she got to Gordon Creek.

The next day we came down to Brigalow Point and she said, "I'd really like a drink." You know, I never had a bloody pannikin with me! I forgot a pannikin! Ah, Christ Almighty, I'd been living too long on my own. I usually did have a pannikin, but I don't know what the hell happened that time. So we got off down at Brigalow Point and I said, "You'll have to lay on your tummy to have a drink." I had a boy with me because we had a spare horse each. I sent the packhorses on and we stopped there for about half an hour.

When she got back on the horse I could see she was pretty bloody stiff, but she could put her foot in the stirrup-iron and get on the saddle all right. We got to Peter Creek and as we trotted down she said, "Oh Charlie, how far is it to Humbert now?" I said, "I didn't like to tell you back there, but it's another five miles." I thought I'd keep her going so I just kept jogging on. She was following me and when we got up on top of the bank on the other side I looked back — and Christ, she was crying. Oh Jesus, I nearly went through the ground. I said, "What's up?" She said, "Oh, I'm that tired, I can't ride any longer."

I felt terrible you know, so I said, "Well, wait a minute."

Then she saw the whitewashed rail I had running from the yard to the house.

She looked and she said, "What are those railings?"

"That's the place," I told her.

Hessie started laughing and crying at the same time then — she reckoned she never ever forgot seeing those white rails. She got down to the house and the next day she was that stiff she could hardly walk.

Well Hessie never did get to Singapore — that fortnight on Humbert turned into six months, and then we went back to Queensland. Hessie had her own car in Ayr and while I was at Woodhouse she used to come out to visit me. Two years later — on April 12th, 1941 — we married, and Hessie came out to Humbert. At least she knew what she was coming to.

Just before Hessie and I were married I built a stone house. Until then I'd only had paperbark huts and these had bark roofs until I got a bit of iron about a year after I first got there.[1] Before then I had no money for iron, and anyway, half the trouble was getting the stuff out.

I'd seen plenty of stone slabs in the bush and I thought: "There's no reason why I shouldn't be able to make a wall with them and finish it off with antbed." I knew antbed was very hard — they used to make tennis courts out of it — but by gees, don't get it wet! It's so slippery then that you can hardly walk across it. It's still hard as the hobs of hell and it only gives a little bit of slip, but *whang!* Over you go.

Once I got the stone walls up I used antbed to finish off

the inside. I'd go out with the donkey team, fill up the wagon with antbed and bring it in and throw it off at the house. I had troughs made from forty-four gallon drums cut lengthways, and we'd smash all the antbed up, put it in the trough, mix it up with water and let it stand overnight. You'd grab a handfull of antbed and slap it against the wall. If you got it at the right "wettage" it worked well, but you can't do all of a wall at once. If you do too much the whole lot will fall off. You apply it up to a certain height and go back again the next day and throw on a bit more. It'll crack a little bit, but you just get more antbed and throw it on and rub it in with your wet hands. You do that about three times and you've got a smooth wall then.

I put up one big room with a bark verandah on the outside, and I ordered some corrugated iron for the roof, but before I could get it out from Katherine the army requisitioned it. I thought Hessie would have to come to a place with a bark roof, but Mrs Martin heard about it and she got Alf to let me have some from the supply at VRD.

In the days before Hessie came we were often very short of water on the place. The lubras used to carry our water up from the river in buckets. We were washing in a tub and pouring water over ourselves, or else we'd go down and swim in the river. Late in the dry there was another problem. The Humbert would get very low and time and again, when early storms made Peter Creek flow, it backed up into Humbert and the backwater would kill all the fish. It was brackish and putrid and that's what we were bloody well drinking! I reckoned a mill would save the lubras having to work so hard, but I still had the problem with putrid water late in the year.

In about 1944 or '45, I called in at Southern Cross in Townsville and told them I wanted windmills to go out to the Northern Territory. They said, "Oh-ho, you're not asking for much are you!"

"Why?"

"Because the war's more important than a station. If we put one on order we might get it in about six months' time."

"Look, I'm in here from the station now, and I've got the truck to take it out."

"Where do you live?"

"I'm way to buggery out in the Never-Never country near the Western Australian border — you've got no way of sending it out. Anyway, what's the bloody mill you've got over there now?"

There were about three mills there already wired up, but they said, "They're already sold and they're to go out to Cloncurry." I argued my case and at last he said: "All right. You can take that one." The company put it on the train for me and I picked it up in Mount Isa. When I got back to Humbert I put it up at the homestead. Roley and I put a well down and struck water at about twenty-five or thirty feet. We stayed it with logs, put the mill over the top and started it going. We waited a long time, but at last the water started to flow. By gees, there was a bit of excitement on Humbert River that day!

The mill had to draw the water up and put it into a 3000 gallon overhead tank. I'd brought out pipes from Townsville and we laid them to the most convenient place for a shower. For a time the lubras still had to cart water up in buckets because I wanted the tank full before we started to use it in the house. Well, we got the tank full and I said, "Right-oh you gins, now you can fill up all your buckets from here." They thought we were mad!

Later on I put up a 30,000 gallon tank at the homestead. I got the real heavy variety which cost me 600 quid. On the evening of the day we started to pump it a dry storm came from Butler Gap — away went the bloody tank, a total wreck!

In 1945 Hessie and I had been on Humbert going on for four years without a family and we'd been talking about adopting a baby. In the meantime Father Dew turned up at VRD. Of course he had a church service there and then he came over and visited us. He said, "I believe you want to adopt a baby?"

I said, "Well yes, we've been after one for the last couple of years now. They tell us that we can't get one."

"Do you mind if I speak to Hessie about it?"

"Of course not. I wouldn't be doing anything about it without her say. She'll be the one that'll cop all the flack." All the women do when they have babies. They're there all the time looking after them and all that sort of thing.

Anyway, I said, "We actually wanted two."

"What age?"

Well I'd always thought of around about two or three years old so they wouldn't be much strain on my wife, but she insisted, "No! The younger they are the better," because she wanted to look after them from birth. We'd discussed it a couple of times and I said, "You're the one that'll be looking after them. Get them if you can."

I told the priest this and he said, "Leave it to me. I think I can fix you up." It turned out that his sister was the Reverend Mother in Saint Vincent's Maternity Hospital in Melbourne.

Some weeks later we came over to Victoria River Downs for the mail and Jack Roden said, "There's a telegram for Hessie she'll be pleased to see."

I took it over to Hessie. "I don't suppose you've won the Melbourne Cup or anything?" I said.

She started to read it and then let out a feminine scream: "Oh my God, we've got a little girl ready for adoption in Melbourne!"

Three days later Hessie caught a plane at VRD that was going down to Alice Springs, Adelaide, and round to Melbourne. Two sisters who had been at the AIM hospital at Victoria River Downs piloted her to the hospital, where she saw Donna. Of course, Hessie thought Donna was "the most gorgeous baby" and all that sort of thing. She said,

"When do I take her?" and the nuns said, "Oh, you can't take her just yet."

I suppose they had to make sure Hessie knew how to look after Donna — although Hessie'd looked after babies before she came to the Territory. There in north Queensland she always had a mob of kids around her. They said, "It's nearly her feed-up time so you can give her a bottle, and when you come back tomorrow you can bathe her."

Well, every time she went down and fed her or bathed her, she wrote and told me, you see. I kept all those letters and put a date on them. I always reckoned that when Donna was twenty-one I'd give them to her, and that's what I eventually did.

It took Hessie about three months to do the round trip because it was back in the war years, and transport had been commandeered by the army. When she set off to come back north she was three times on a plane in Melbourne and a damn officer came along and pulled her off. Eventually she took the train up to Ayr, where she had a sister. She stopped there for three or four nights and then caught the train out west to Mount Isa.

That was when I ran the gauntlet of the military. They had armed personnel all along the road, stopping anyone from going backwards and forwards. You had to give a specific reason why you were going into Mount Isa or wherever it was. They didn't want women in the Territory for fear of Japanese invasion those days, so I knew they'd block me if they got the opportunity.

I went from Humbert across to Newcastle Waters and down to Tennant Creek all right, then along the backtracks from Tennant Creek to Rockhampton Downs. From Rockhampton I went through to Alroy Downs, and from Alroy down to Avon. By that time I was nearly to Camooweal and people from the stations around there were allowed to go in and out of town. I managed to get a wire away from Camooweal advising Hessie that I'd make it to Mount Isa about five o'clock. I got in about two minutes to five. Then we came back the same way and got past the army no trouble.

When Donna was nearly four we got Betty who was about thirteen months old. She came from the convent in Brisbane. When Hessie brought Betty back she picked up a new car I'd bought for her in Mount Isa and motored out to Elliott, and I went with the Lend-Lease truck to meet her there. I got there at about midnight but Tommy and Eileen Thomson and Hessie were still awake, so I went in to have a look at Betty. She had a mosquito net put over her you know, and she was just lying with all her hair over the pillow. By gees, she was one of the prettiest little kids I'd ever seen.

We got governesses to teach them on the station, and their school work was done by correspondence. Then when Betty was eleven and Donna thirteen, they went to a boarding school in Adelaide. There's no mistake about those South Australian correspondence lessons. It was really outstanding, so much so that when the kids went to the Convent in Adelaide they were so far advanced that they were jumped to another class. They could well and truly hold their own.

We taught Betty and Donna to ride when they were young. When yarding up cattle I'd hold one or the other of them in front of me on the saddle. I used to take up Donna till Betty woke up that she couldn't go for a ride and started to put on a bloody turn. Then I got Roley Bowery to hold either Betty or Donna while we yarded up the cattle, and they'd kick and squeal, and squeal and kick, yelling out all the time. Talk about being excited!

They learnt to shoot when they were about six or seven years old. They had a .22 rifle and were mostly on target practice, but Donna used to clean up the crows because they were stealing our eggs hand-over-fist.

Dingoes — noted that they are worse now than they have ever been. He gets about 80 scalps a year by poisoning and shooting. Appar-

ently he makes a regular set at them, and has built a platform in a tree out near his killing yard, and often sits up there of a nite to get a shot at one.

F.H. Bauer, 1957

I had a set-up there for shooting dingoes. We'd knock the killers under a big tree out on the flat, not far from the house, and dingoes used to come up scavenging for the bones. I'd driven some iron bolts into a tree for a ladder and I'd go out in the evening about the second day after butchering, and climb up. When the dingoes came you'd be looking straight down at them. If you keep quiet, on a still night you can hear a dingo trotting about twenty or thirty yards away.

At the start I used to wait till Betty went to bed, then I'd take Donna out with me. Later on I used to take them both up. Donna and I would go out just before sunset and Hessie used to lay down the law to me: "Now Charlie, you have her back by nine o'clock. Look out she doesn't go to sleep when she's in the tree."

I'd leave Donna and the empty shotgun down on the ground while I climbed up with a big hessian feedbag. I'd get up about ten or twelve feet and tie the bag up there with a strap, round and round. Then I'd put a packsaddle surcingle strap down — they're about twelve feet long — and Donna would tie the shotgun on, and I'd pull it up and put it in the bag. Last of all I'd lower the strap again and say, "Now you just put it around your waist." Donna was only light and I'd pull her up and I'd put her straight in the bag so if she slipped or went to sleep, she couldn't fall. Mosquitoes were fairly bad, but when she was in the bag she had a lot of protection, and it didn't matter if she went to sleep or what the hell she did. I'd say, "You keep quiet and don't you start talking." Later on I'd have Betty and Donna up there together, and by gee they were good kids. There wasn't a peep out of them you know, even when a dingo eventually came along.

They loved life on the station. Betty was in America when Hessie wrote and told her that we were selling the place. She said, "Mum, if I'd known Dad was going to sell Humbert I would have flown home straight away and blocked him." Oh yes, they had their horses there and they were always out mustering. By Jesus they're great mates — they give you the shits! When they were in Darwin you know, if I rang Donna and she wasn't home, I'd bet she was around at Betty's and vice versa. Now one's living in South Australia and the other in Queensland, but they're still all the time ringing each other up.

> *Mitzi Downs: Aged 4 years at removal. This female Aboriginal child was removed from the native camp at Victoria River Downs by Protector Pott ... It is once instance where, unfortunately, the full consent of the mother was not obtained by the person removing the child and some distress was caused.*
>
> The Secretary, Department of Territories, Canberra, 25 January, 1952

A law came in that all the half-caste kids had to be taken away from their mothers and taken up to Darwin. Well you never heard of such a ridiculous law in all your life. It was the saddest thing I ever saw — little kids dragged away from their mothers. You picture, any of you, anyone that's got family now, a policeman coming along when your kids are only six and seven years old and dragging them away from you. Can you imagine how they'd scream, and the turn they'd put on when they're taken away from their mother?

I was involved in such a case at VRD one day when I came over to catch the mail plane. That particular plane was flying straight through to Darwin and when I was on the airstrip I saw these two kids there, hanging to their mothers. Their blackfella fathers were there too.

At last the plane was ready to go. The pilot sat up in the seat and I hopped in, and the next minute a policeman was there, a fella by the name of Jack Potts. He brought the kids over to the plane and they took fright then. He grabbed one and when he went to put it in, it hung onto its mother. I admired the lubra. She turned round and she grabbed one and she kept on saying to him, "Don't you worry, you're all right, you go to Darwin and learn lessons." You could see it was against her will that she was doing all this.

I was sitting up in the middle of the plane and didn't want anything to do with it. The pilot didn't want to get involved either. He just sat there all the time while this shivoo was going on. The mother grabbed her kid and got up in the plane. Potts brought in the other one, a little boy, but when the little girl's mother went to jump out of the plane the kids followed her down, and went to jump out too. She turned round to me and said, "You hold'im boss, you hold'im." God, I didn't want anything to do with it. It had nothing to do with me. Then the little girl took a half-hitch around her mother's hair. She got one finger wound around it, and I'm damned if the mother could get away from her.

Christ, I'll never forget that until my dying day. I said to Potts, "Why don't you let the poor little buggers alone? Let them go!" He said, "Oh I might lose me job." "Yeah," I said, "your job'll be all right. There'd be too big a stink go up if you did your bloody job over that!" I don't know whether he just wanted to be a dinkum policeman or what — the bastard was howling himself! I'll never forget that you know, crying him-bloody-self! And if I wasn't crying I wasn't far from it.

Anyway, at last, after a lot of pushing and shoving, the mother jumped off the plane, and the pilot started it up. But God strewth! — I'll never forget the way those kids kept on looking through the windows or back down the stairs. They were both howling of course, and so were both the blackfella fathers and mothers. Don't let anyone tell you they haven't got feelings — of course they've got bloody feelings!

The plane eventually got off the ground and the kids looked around and I was the only one they knew. Apparently they knew me by sight from my coming over from

Humbert to get the mail, and they fastened onto me then. One sat on each side of me because I was somebody they knew, though I'd never spoken to them. Well, we took off and when we got over Jasper Gorge they were still crying and looking back through the window towards VRD but before we got to Timber Creek they both flaked out. I suppose it was the drone of the plane that sent them to sleep. The little girl ended up going to sleep on my knee and I just let her sit there, poor little beggar. The little boy went to sleep on the seat alongside of me.

When we got into Darwin, darn me, they woke up. They jumped up and looked around, wild as brumbies you know. They were looking at me all the time, and they were looking out through the window back towards VRD again, their home. At last we landed and Ron Ryan was one of the Welfare mob that came to pick them up. I forget who the other fella was, but they were both strangers to these kids. I thought, "Here's gonna be a bloody go again now."

As soon as they got off the plane, of course, they claimed me. I said to them, "No, you'll have to go with this man now and he'll look after yer," but their little minds were shattered by this time. They didn't want to leave me at all. At last Ron grabbed hold of one kid and carried it, and the other bloke carried the other. I could see them going out in the car and as far as they could they were looking back at me.

God, I'll never forget that. If ever there was any bloody blackbirding — and this was our bloody "John Government" for you — this is what they were doing. These are the bloody things that a superior force will do. Oh Christ, I'll never forget that as long as I bloody well live.

My wife looked after a half-caste lad called Les. We'd been educating him and they decided to let him stop there. Well! It's marvellous how this news got around. In no time the station lubras at the Humbert were telling us, "Oh, so-and-so want to send'im over her half-caste kids. She said you can have 'em." Here they were giving their half-caste kids away! As long as they weren't going away to Darwin they could stop at Humbert with Les and go to school there. They couldn't make out why we didn't want them all.

Eventually Hessie reared about four or five half-castes I suppose. There was Larry Johns, Rosie Gordon, Les Humbert and one or two others. Larry Johns came from the Timber Creek camp and we got Rosie Gordon out of the Gordon Creek camp on VRD. Her father was a bloody brainless bastard. A sort of ringer, big and tall, but no bloody brains about him. Les Humbert was about four years old when Hessie came out to Humbert. His father was Roley Bowery and his mother's name was Milly.

Milly was about seventeen or eighteen I suppose when she died over opposite Ivnors Pocket. I was away at the time in Queensland and a bloke called Bill Grogan was fencing on Humbert with old Humbert Jack. Jack ran away from Bill and took off down the Wickham with Milly. I don't know whether she had Les with her or left him with her grandmother, but Milly was pregnant and she died in childbirth. She had the kid and the Aboriginals reckon the poor little bugger lived for two or three days. Of course, it starved to death. That's the mentality of the blackfella again. Anyone else would have said, "We'll take him back to the station." It could have been raised on goats' milk you see, but I believe it cried and cried and cried, all the bloody time, and they just let it die.

When Les Humbert was about twenty we were mustering Ivnors Pocket and we came on about eighteen or twenty head of cattle. Ivnors Pocket is naturally wild country and very bad going in there, a maze of creeks and gullies. If you came onto cattle in that country they'd take off, and if you lost a horse you'd have to run the bugger down to get him again — even an old quiet broken-in horse.

There was a good flat for a couple of hundred yards right around and we all took off after these cattle. The next minute I saw Les hit a bloody tree. He went straight up like a gunshot and *bang!* As I galloped past I looked and he never kicked. I knew then he was knocked cold. I circled around and came on back to him and jumped off my horse. I ran over to him and he was just lying there as though he was dead. I think someone else stopped his horse.

From what I could gather he'd jerked his horses' mouth. It chucked its head in the air and didn't look what it was doing, and cannoned straight into the tree. Where he hit the bloodwood you could see the mark of his knee and the stirrup iron right around, and his head too. Well that buggered up our mustering, so I said, "Righto, let those cattle go."

Les was out to it for about twenty minutes or half an hour, and I was wondering how the bloody hell I was going to get him across the Wickham River. I got him on his horse eventually and he kept saying to me, "I'm spoilin' your muster, I'm buggerin' your muster up aren't I?" I said, "All you have to do is just sit on that bloody horse and keep quiet and don't talk." Then he'd say again, "But I'm buggerin' your muster up." He was a bit skewy you know, and he kept on saying to me, "Ah, me head's spinnin' and me eyes are flickerin'." That was concussion you see.

We got him across the river and on about another four miles up to the camp and straight onto his swag. Then I sent somebody in with a letter to Hessie. "Les hit a tree, suffering from concussion. If you can get out tonight, come on out with the truck and bring a couple of lubras to give you a hand." We got him in to the station that night and he was laid up for about eight days.

All these kids were reared and went to school at Humbert, taught either by my wife or a governess. Larry and Les were both good lads and when they got older they went into the stockcamp. When Les Humbert left us he had about 1200 or 1400 quid in his account. Larry Johns was younger than Les, and he had about 800 or 900 quid. I'd give them their cheques monthly and Hessie would say to them, "Now you bank half of that and you can spend the other half." She used to make them write to the bank and send the bankbook down themselves, to teach them what to do.

They turned out pretty smart men, reliable if they had a job to do, and as time went on they married coloured girls. Larry got a good job in Katherine and Les has been overseer at the Kidman Springs Research Station for years.

11 The End of Isolation

In the beginning Humbert was a hell of an isolated place and we did just about all our travelling on horseback. There was a monthly packhorse mail in the dry times and in the wet it was every six weeks or couple of months. Then at the beginning of the thirties I think they were starting to get the odd mail truck coming out. Of course, in those days the roads were against using motor vehicles.

Most stores came by boat to the Timber Creek Depot and from there to the stations by donkey team. To get the rations from the head station to the outcamps, VRD used camels. Other places used wagons pulled by donkeys, horses or mules, or down near the desert some stations used camels. And of course in the wet season it was a hell of a lot worse. About the only way anyone could get around then was on horseback.

I never used camels myself. VRD would send a couple of boys — generally a boy and lubra — up to, say, Mount Sanford with the six-weekly supply of rations, and when they came back again they took a load out to Pigeon Hole. There was no road suitable for a truck. It was best to send a damn camel team because you knew it would get through. Montejinni and Moolooloo had about three camels each to shift the loading over there. They stopped using camels about the end of the thirties; they turned them loose then and some of their descendants are running on Humbert to this day.

Talking of camels, Wave Hill and a few of the other stations used to get some of their stores brought up from Alice Springs by the Afghan camel men. A lot of blokes had a set against Wave Hill those days, because Vesteys were very poor payers as regards wages, and they used to skimp their tucker bill and all that sort of thing. It was close to the end of the year and the Afghans were bringing the Christmas supplies from south. Of course, they wouldn't touch any pig meat, but they often carried it unwittingly because the station would arrange to have it packed up as something else.

This particular time the bloke who had the set on Wave Hill hopped out and met the Afghans when they were still about sixty or eighty miles away in the desert. He had a cup of tea with them and then he said, "Will you sell me some ham?"

"Oh, no! Our God say can't, our God say can't."

"Pork or ham, anything? You got any pork?"

"No!"

"Oh, you've got it there all right. It's in those cases there. I know for sure because I've got a mate down south who says that's how they're sending it up to Wave Hill."

Well, the bloody 'Ghans wouldn't touch those cases again. Wave Hill had to come all the way out with their own camels and lift the stores, and I believe in that particular loading there was no ham at all!

> ... *my horse losses are terrific, and this year I have lost between seventy and eighty head, through the Kimberley or Walkabout disease — forty odd of these being broken in — a tragic loss to the station. This has been the heaviest loss I have had to date ...*
>
> Charlie Schultz, 26 December 1947

Jackie Lewis and Dave Ebzery were a couple of young fellas who came out from Ayr to Humbert with horses in 1934. We were always needing horses on Humbert because they were always dying from Walkabout disease. That knocked hell out of our horses. We might lose four or five at once, and it was heartbreaking. For years no one knew what caused it, but eventually they found out it's caused by a poison plant, a rattlepod called Crotalarium.

Jack and Dave brought over about thirty head of horses and Jack stopped for a couple of years. I didn't have money to pay him so I got him a job on VRD. Late one year he came over to Humbert and said, "Are you goin' home for Christmas?"

"Yes, I'm thinking about it."

"Right," he said, "I'll go with yer."

Well the way we used to get home was to go by packhorse into Katherine, catch the weekly train to Darwin, and then take the boat around to Townsville.

We left Humbert on November 9th and were in rain practically all the way in. We had to swim our horses over the Victoria River at Dashwood Crossing, and to get our gear over we made a raft out of a big tarpaulin and pack saddles. We laid the tarpaulin out on the ground, put our pack saddles in the middle, wrapped the sides of the fly over the top and then tied it up with ropes. One of us swam in the lead pulling it with a rope and the other got behind and pushed. You could carry anything up to a ton and a half like that, no trouble at all.

We came across and camped on the King River. I think we might have got in round about four o'clock and I said to Jack, "Oh well, we'll try to get across to the other side so's we can hit Katherine tomorrow." We had to float our swags across there too. Next morning when we were packing up, Jack shook his swag and a little green centipede fell out. Now it was only a few inches long, but Jack bounced out of the bloody road and it ran over to my foot. I just stopped still, thinking it was going to run *over* my foot, but the bastard bit me!

I can still hear Jack laughing you know. Yeah, laughing his bloody head off! And he said: "Why didn't you shift?"

I said, "Look, a hundred bloody centipedes or snakes could come along and if you keep quiet the bastards will just crawl over your bloody foot." He shook his head and said, "Well that's the best piece I ever heard of, letting a centipede bite yer."

Anyway, we were throwing the saddles on and I pressed myself in the groin. Talk about pain! This dropped me. By bloody Christ it downed me! When I came good we finished what we were doing there and about twenty minutes after, he said to me: "How do you feel now?"

"Not too bad. By gees, I never want to go through that again."

"Where did you get the pain?"

"Here." I pressed the place and Ho! Jesus! Down I went again!

What had me worried then was whether I could get on my horse all right. Jack said, "There's a log over here," so he got me alongside and I sneaked into the saddle. We were still twenty-five miles from Katherine and I was wondering whether I was going to be able to ride. I was all right, but as I rode the horse along for the next couple of hours, the bugger kept on saying to me, "How are yer now Chas? You going to press it? You going to press it?" You know, he was a lot younger than me, Jack. Anyway, after a while I got cheeky and I did feel around there, and the pain had gone. It was just one of those little green centipedes.

Before the end of the day, "Leaping Lena" had grown into "Limping Lena." Something went wrong with the works and she had ignominiously to be pushed part of the way home by man-power.

The Northern Territory Times, 27 January 1931

We got to Katherine but we missed the weekly train to Darwin. Oh God! That was little "Leaping Lena". It was only a three-foot gauge and by God it was slow. It only went as far as Pine Creek the first night. A week later we got the next train to Darwin and then we found we'd missed the boat by three or four days. The boats were monthly, taking passengers and mail back and forth between the Dutch East Indies and the southern states. We had to stop a month in Darwin and finally got home for Christmas on the 11th of January.

The next time I went home for Christmas I got there about the fourteen or fifteenth of January. In those days you didn't know when the trains or boats were running.

You were out bush all the time and it was only once in a blue moon you ever came to town, so you took pot luck that you'd catch the train or boat shortly before it left, rather than arrive shortly afterwards.

Have had very big floods in our Rivers ... trees that were 50 to 60 feet high on the banks are now standing upright in the centre of the stream.

Alf Martin, 25 March 1935

... almost every tree is uprooted or swept away for 1/4 mile lengths where the full force caught them, and there were some very big old trees ...

Sister V. Stewart, 19 January 1937

Back in the late 1930s there was a hell of a big flood in the Wickham River. Down at the Humbert junction the water came out a mile and a half to two miles either side of the river and it was the best thing that could've ever happened. It washed out a hell of a lot of the trees, and as you know, vegetation like that keeps all the ground together — there's no erosion. Once the trees went there was a certain amount of erosion, to the extent that at least you could get cattle over the river in places.

Before this flood you couldn't cross damn cattle over the Wickham on account of the steep banks and the vegetation on the bottom of the river! And there were so many trees along the banks that you couldn't see from one side to the other. Beautiful big trees they were — paperbark and leichhardt pine and fig trees. It was only odd places you could cross cattle over.

At the end of 1933 Roley Bowery and I went down to Adelaide for a bit of a break and when we were down there I decided to buy a Chevrolet ute. We had our swags and everything with us so we motored back in the Chev. There wasn't a skerrick of bitumen once you got past Port

Augusta, and we struck heavy rain. Rain fell from Newcastle Waters right down through Alice Springs and as far south as Port Augusta.

We were in rain right up to Newcastle Waters and when we got there the floodwaters in Newcastle Creek were just starting to recede. Water about four feet deep was still on the road, but a Department of Interior mechanic at Newcastle said, "We can get a mob of Aboriginals and push you through, providing you keep on the road." We got over that way and I remember we were that pleased to get going we went right through to the Murranji.

The rain slackened off once we got onto the Murranji Track so we camped that night at what they call the "jump-up". When we got to Peter Creek we had to cut a track through the stones and push them down, and with the Aboriginals pushing we managed to get over. That was the first vehicle that got right to Humbert River homestead.

As time went on we made crossings on Humbert River and Steep Creek. There were no graders or tractors in those days, but we usually had a team of Aboriginals to give us a hand to push. I'll never forget the day we first crossed Steep Creek because it was all washed out, and we had to use picks and shovels to make the crossing. There was no vehicle track in to VRD; we made our own road by driving across country.

> *(Aeroplane should arrive at V.R.D.) Aeroplane arrived at about five in the evening.*
>
> Humbert River Station Diary, 16 June 1929

I was there when the very first plane landed at VRD back in 1929.[1] I came over with all the blacks from Humbert to see it. There was a flat right along there so we all got out and were sitting up where the stockyards are now. It was the best part of half an hour late too. The sun was getting down and we knew it was coming from Wave Hill. Of course, about nine-tenths of those waiting were VRD blacks. Everyone was waiting and waiting, and a Humbert gin called Fannie was the first to see it coming. Fannie only had one eye, but she looked up and said, "Oh, I can see'em somejing longa sky!"

A couple of gins there, oh, didn't they let out a squeal, and somebody said, "Look, those gins reckon they can see something." We were all stargazing round — it was just a little speck you know. Well, it came in and landed on the road and all the blacks stood back. They were a bit frightened of it at first, but they saw the whites there and realised it was safe.

It was on August 8th, 1939, that Eddie Connellan flying a single-engined Percival Gull made the first flight of 900 miles to Wyndham, calling at Mt. Doreen Station, Granites, Tanami, Inverway, V.R.D., and Wyndham.

Hoofs and Horns, November 1959:59

About ten years after the first plane came to VRD, the first commercial plane service started. Eddie Connellan had been thinking of starting an outback mail service — I remember when he first came down to VRD to talk to people about it.

Well, Eddie got his air service going and we started to get all kinds of things sent out by air — not just mail. Until then we had to get all our perishables in by truck. We thought it was amazing to get out things like pork chops or fish, or a leg of lamb, and to get it still bloody-well frozen too. Oh, we were living like kings! It was pretty expensive, but the fact remains, if you want it you'll pay anything.

The Second World War really helped open up the Territory. For a start, some of the roads were improved, and after the war ended there were cheap motor vehicles and machinery available. The Americans put down what they called the "Ninety Day Road" from Mount Isa across to the Threeways. That was mighty work. They did it in three sections and finished it in three months. The road was a bit on the narrow side, but it did the job.

Cattle prices went up too, but Vesteys got a monopoly on contracts to supply beef to the army, and they sent a manager out to buy cattle at whatever price they offered.[2] That was in '42 or '43. Vesteys were paying VRD £6 or £8 a head for their bullocks and were getting about £8 or £10 themselves for them. There was a hell of a squeal on for a while, but nothing could be done about it.

I was a bit lucky. Len Mann was managing Wave Hill at the time — he came to the Victoria River country in 1934, and was a good man on a horse. Len was sent across to Humbert to buy my bullocks and he said, "I'll give you £3-10-0 a head all round before I look at them." He offered me less than the VRD price because my cattle were younger and lighter.

As soon as he said that to me I could have kissed him you know, because I knew that price would clear me with the banks and leave me with about twelve or fifteen hundred pounds to carry on with. But when anyone worked like that towards me I generally woke up there was a catch in it somewhere, and the cattle were worth a darn sight more than what was being offered.

I said, "Are these supposed to be station delivery?"

"That's right."

"That means you'll pay me to take them across to Wave Hill?"

That would mean another week's work for me and would have cleared the wages of the blokes I had helping me.

"Oh no, it's station delivery on Wave Hill, but I can't pay yer for that."

If I'd accepted that offer, Len would have cut out the rejects on Wave Hill — we'd have had to deliver the entire herd first and then take the rejects home again. We'd be doing the droving for Wave Hill but wouldn't be getting paid for it. I said, "No, you give me £4 a head on the place and take your rejects out." Len said, "Well, all right," so we rode down and started culling them.

When I gave him a look through them I was riding with him and he was cutting out the small ones, naturally, so every time I saw a small one up ahead I'd say, "That's a lovely bullock over there Len — this one's not a bad bullock, he's a beauty," and I'd switch him away from the little fella, you see. I did that to him on seven or eight occasions, but he did see one or two small ones here and there — I couldn't do the lot! I let him go about eight or nine over a five per cent cut because, like I said, the price I was getting was all right. Then I said, "Well, that fifty's up, Len."

"Oh, is it?"

"Yes," I said, "we'll go over and count."

So we went over and counted and there were about eight or nine too many. Len looked at me — he didn't know whether I was going to say to him, "I'll put ten back," or not — but I let it go at that. We walked down to the camp then, and I didn't want Len to hang around and count the cattle, so I said, "The sun's getting down isn't it?" He said, "Yes, and I have to get to Wave Hill tonight." The roads were absolutely frightful back then, and he had to go back via Moolooloo and Montejinni. "By gees," I said, "I can see you getting there about eleven or twelve, or one or two in the morning. That's on condition you can get off Humbert before dark. Then the travelling won't be so bad."

He said, "Jesus, it's going to be just about dark when I get off!"

"Well," I said to him, "what do you think?"

"How many should you have there now?"

I'd had about 1000 in the mob *before* he'd taken the rejects out, but I didn't tell him that. I said, "About 1000, Len."

That satisfied him and away he went. As soon as he got out of sight, the bloody sixty-odd that we'd cut out I shot back into the mob again. They weren't going to pay me for taking them from Humbert so I evened up on him that way. That's when I knew that at last, after all those years, I was cleared of the banks. By gees, I think that was about one of the happiest days of my life.

Military truck with Sergeants Taylor, Grant and Private Andrews of 6th Division A.I.F. (overseas) arrived 6pm to instruct throughout country in the use of Tommy Guns, Bren Guns and hand grenades.

Timber Creek Police Journal, 24 June 1942

Warning received from Sgt. In Charge Observers Unit at Depot stating that word had come thro from Hd. Qrs. that it is expected the Japanese will make a raid on Timber Creek by planes. Slit trenches being dug at the Depot.

Timber Creek Police Journal, 15 November 1942

During the war the army asked VRD to send word out to all the stations that they wanted the station owners or managers to come in for a meeting. The meeting was to last for about ten days and we were to bring our own swags. VRD would supply all the tucker. It was a wonderful set-up they sent out. They really thought the Japs were going to land and they gave us the world, as the saying is.

There were about twenty-five of us and we camped down there at the old hospital on the Wickham. Those days the hospital was empty. Three army blokes came out for the meeting — one was a captain, I think another bloke was a sergeant-major, and the third was a corporal. They gave us lectures about different things and they reckoned that if the Japs came in overland they'd let a certain amount of them go through, then cut them off. They'd be short of petrol and supplies for a start. The Captain reckoned that if the Japs landed, the Australian army would have about 200 hit-and-run soldiers stationed in the VRD and Humbert ranges to attack the Timber Creek road.

On account of me being a station owner they made me a sergeant and I had to show them where the big springs were in the back country on Humbert and Bullita, places where camps for guerilla fighters could be set up. One spring I picked out was at the head of Figtree Creek. There's a couple of beautiful springs there right in the hills in amongst timber — crystal spring water. Then they wanted to know if there were any cattle around there that would provide beef for them. Of course being the war years it'd be catch as catch can — whatever you saw you'd just blow it over. As it turned out the army only ever stationed men at the Depot at Timber Creek. They had a couple of army blokes on guard right on the river, watching to see what was going on.[3] Otherwise the Japanese could have come along the Victoria River by boat and no one would have been any the wiser.

Now, they had Jasper Gorge mined and they were going to use it as a trap. The Captain reckoned that when the Japs landed they always sent someone on a motorbike on a reconnaissance first, to see where the roads went. Our side wasn't going to touch him unless the others were following close behind — the plan was to let him go back and report. The Captain said no doubt the Japs would be coming up in motor vehicles and when the first lot came through, we were going to blow the top of the gorge and the rocks would bury the road where it runs right in under the cliff. What that didn't wreck wouldn't be able to go any further and they were also going to get in behind them so they couldn't turn back.

On the other side of the road is Jasper Creek. They gave me to understand that when they blew the cliff up, some of the surviving Japs would probably hop into the water to swim to the other side. Well the plan was to have some of our men on the other side and as the Japs were swimming or coming up on the bank, they'd be shot. I don't know whether they ever shifted it, but there was dynamite in the gorge for at least eight or ten years after, you know. It was still in boxes up there and never used.

Into the hands of these barely trained men were put powerful armouries. Typical were those at Victoria River Downs, stocked

> with 24 rifles, two Lewis machine guns, two Owen sub-machine guns, a large stock of ammunition and explosives for demolitions ... But the army lacked control over this amorphous force.
>
> R.A. Hall, 1989:130

The army sent out different military weapons and they gave us about ten days or a fortnight's know-how about all these guns. We had a go with .303 rifles and machine guns, and they had a bit of dynamite with them too because they showed us how to set it off. Actually a lot of bush people already knew how to do that. I always remember a ninety-second fuse they had. I still don't know how the hell they could work that one out.

They had anti-tank guns there too. There were several blokes practising with anti-tank guns, shooting at a big log about 150 yards across the other side of the river. That's when I found out that I couldn't stand the recoil and noise of a big gun. It had never occurred to me that the report would affect me, but Jesus, I just couldn't stand it. I had to put my fingers in my ears and get right out of it. It's a strange thing, but before this ever happened my wife would sometimes say to me, "If you were in a war, Charlie, you'd never be able to stand up to it."

They had the anti-tank guns, and they had grenades. We were all given half a dozen hand grenades to learn how to throw them and to get the feel of them. All of us had a crack at throwing one. We went down behind the bank of the river there, right on the waters' edge, and were throwing them over into a gully. Somebody gave one to old Gus Anderson to throw. Gus was a Scandinavian who worked on VRD as a handyman. He could do anything, Gus, you name it he could do it — anything except throw a bloody hand grenade!

Christ, he didn't throw it ten feet! Boy oh boy, was there a scatter among us. I joined a mob that took to the water, boots, leggings and all. Everyone was holed up in gullies or in the waterhole and poking their heads up and asking, "Has it gone bang yet?"

"No!"

Down we'd go again, ducking our heads. One of the blokes with elastic-side boots on bloody near drowned! Oh Jesus, he only just got out of it by grabbing hold of a log in the water there, and he said, "It's better to get your head blown off, rather than bloody well getting slowly drowned to death." As luck happened the grenade didn't go off.

> *We do not know what will happen out this way but at the time of writing things are very much out of order. Eleven men have joined up from Victoria River Downs ... There are only four stockmen left on Victoria River Downs and we think that at the first bomb dropped in this district all the blacks will go bush.*
>
> Alf Martin, 9 March 1942

Most of the time the war didn't cause any real labour problems on the stations. Sometimes the stations might be short of a white stockman because they'd been called up and gone to war, but those days there always seemed to be enough blackfellas around, and anyway, in the station camps the only white men you had were the head stockman, the cook and an offsider. At odd times they were short of a cook and several places had blackfella cooks. That was unusual, because they're inclined to be on the dirty side. They weren't taught hygiene right from the word go. Actually the lubras were better than the boys, but by gees you had to watch them too.

The police demanded that you have another white man with the head stockman in case of trouble with the Aboriginals. They reckoned you needed a third white man because the cooks were usually broken-down old chooks, and the camp kitchen was kept well away from the dust, anything up to half a mile away from where the head stockman was working in the yard. They were always on my back about getting another white man on the place, but economics was against it — I just couldn't afford to pay anyone.

Those particular times we were still carrying revolvers

in case of trouble with the blacks. In the mustering camp even the bloody old cook had a revolver. You took no notice of men wearing guns because all the whites did it you see. That was before the old Aboriginal got spoilt. He was a big asset then. They reckon he hasn't got much brains, but he was cunning enough to know that if there was a war on he had to do his bit, or he was going to get blown over by the enemy sooner or later.

For a time there was a real fear that the Japs might invade, so the army went around and armed all the local stations with .303 rifles and crates of ammunition. One of the policemen stationed in the district during the war told me an amusing story about arming the stations, but I have to point out that this particular bloke was the biggest bloody exaggerating liar under the sun.

According to the cop, at one of the stations the manager was a full-on communist — he had a big red hammer and sickle painted on the door of his store. When the army blokes turned up at his place with the guns, the officer went up to the manager and said, "Mr ... , in the event of the Japs landing I've bought yer some rifles and ammo."

"Oh did yer?"

"Yeah, they might come in handy. Where will I put them?"

"I don't care a damn where you put them. I don't want them. Put them outside under a tree over there or anywhere you like."

"But Mr ... , these are in the event of an invasion — don't forget there's a bloody war on!"

"I don't care a bugger," he said, "if there's a half a dozen wars on. I don't want these things here."

Well the officer was a bit shocked and he asked, "What would you do if the Japs did turn up here?" And the manager replied, "The same as anyone else would do. I'd ask them to come in and have a cup of tea." Of course that was too much for these Aussie soldiers. They were all set to lock this bloke up, but the local cop calmed them down and reassured them that the manager was really a harmless bastard. Yes, you see some bloody hard cases in the Territory all right!

Charlie Schultz now has an airstrip at Humbert River Station.

Hoofs and Horns, November 1958:57

We didn't have an airstrip on Humbert until about 1958; for years before then we used to have to go over to VRD for our mail. Then I went over one time and noticed one of my boys in the VRD blacks' camp. I said to Jack Quirk, "You've got one of my boys sitting down in your camp. Would it be okay if I get him?" I thought he said yes. Anyway, later on when I was just about to go I said, "I'll get that boy now."

"You'll what!"

The way he reacted you could have knocked me over with a straw! I said, "I'll get my boy there now."

"If he wants to stop here he'll stop. Don't forget I fed him."

Quirk reckoned he fed him and looked after his mother and Christ knows what for years and years.

We had a blazing row about it and that's when I told him, "You're known over in the west as 'Stand-over Jack' and 'Bulldozer Quirk', you bastard." By gees, it's a wonder he didn't clock me you know. We walked from the Big House down to the office, and he was bloody-well shivering and shaking all over — by Jesus we were going! He ended up saying to me, "Don't grace these bloody steps again."

"What do you mean?"

"Don't come into this bloody post office again."

I said, "Don't forget this post office is Government, old fella. Don't try and put it over me the same as you've been doing to your bloody workers all these bloody years."

That's about as far as it went, but the rumours that got around the country — oh Christ! It went all round the station that I offered him out, but that definitely wasn't right — I did nothing of the sort. I was always inclined to be a bit quick-tempered though, and after this run-in with Quirk I decided I'd had enough of having to go in to VRD for my mail and deal with the likes of him.

I went back to the Humbert and said to Hessie, "I'm gonna put down an airstrip." She threw her hands in the air in holy horror and said, "Where are you going to put one down here? Not over on the Wickham River?"

I said, "We don't want to be running over to the Wickham River ten miles away." A lot of the stations had their airstrip out five, six, and eight miles. Most of them could've got one closer, but they had no commonsense. "We'll put ours on the other side of Peter Creek."

That was less than half a mile from the homestead, but Hessie said, "Charlie, what about those trees?"

"We'll just have to grub the buggers out, but I've had it with VRD and bloody old Quirk."

I was always independent and hated putting anyone to any trouble or being messed around myself — "I'm gonna go up there this afternoon and put in the rest of the day grubbing trees where I think we could put the 'drome down. I don't know how long it'll take us, but there's only one way to find out and that's to just go out there with the blacks and have a go."

She said, "My God, I only hope you're not going mad."

"I'll tell yer tonight when I come in."

We got stuck into it just before two o'clock. I had five or six sets of blokes going and each man had either a crowbar, a shovel or an axe. I had to peg the damn thing out for a start so I started a few of them on a couple of big trees there, and I went straight ahead then and pegged it out for about a quarter of a mile. I wanted to see how much could be done that day — God strewth, I think we did about 200 yards!

I learnt from VRD's experience. VRD put a strip down with bulldozers and I never saw so many suckers come up in all my life. You could no more get a plane down there than sail a boat on it, so we didn't just cut the trees off, we grubbed them out. I said, "We're grubbing all the trees out, roots and everything." We had to dig down and shovel the dirt out, and we kept the dirt handy because we had to fill the holes up again.

Hessie came over about four o'clock with the smoko and she said, "By gee, you're getting things going all right!"

I said, "We'll have an airstrip down here. The timber is a bit denser down further but that'll make no difference. I can see now what we can do."

Well, we grubbed out 1627 trees and had them down and cleared in six weeks. There were a lot of big logs which we cut into lengths — I never saw so much firewood around a station before. We scooped up the branches and leaves and burnt them, and then we had to pick up the bloody chips. That was our downfall you know. If we'd had a crosscut saw or chainsaw we could've cut up the timber and not had any chips to worry about. I thought the job would never end — I still have blisters!

Before they let us open the strip they sent out a plane to inspect it, and the bloody thing had to have a puncture. It was a slow leak and they sent word back later saying, "Check your airstrip again. Appears the plane picked up a puncture." We went and tracked the plane back and found a little bit of a spike where somebody'd chopped off a conkerberry bush, but they made us wait for about another bloody month before they opened the strip. The first plane to land once it was officially opened brought my mother out for a holiday. It was the best day's work that Quirk could've ever done to try and stop me getting my boy that day.

That airstrip going in just about ended the isolation of Humbert River, although the roads were still a bit rough in places. If anything went wrong, someone got sick or we wanted to get some place in a hurry, we could get out on a plane in no time.

As the roads improved, the motor vehicles got more and more reliable too. After the Lend-Lease truck we got others, until at the last of it I was getting big nine-ton Toyotas that I made into cattle trucks. I had a total of about thirty-five or thirty-six Toyotas altogether. I'd trade them in every second year and they'd allow me so much on them.

12 Rough Hewn Men

> *He was a rough-hewn man, moody and aggressive, somewhere between fifty and sixty years of age, who used to delight in giving me the horrors at the dinner table ... he would glare at me and his great monster of a clenched fist would come down with a crash on the rough wooden table like an exploding bomb, lifting the knives and forks a foot high and scattering them. He was a bully and at times rather sadistic.*
>
> Elsa Chauvel, 1973:134–35

About Wason Byers, I can only go on what I heard of course. He was a bastard of a man — a big rough sort of a diamond who used to play on the way that his voice was so loud, you know. Everyone more or less hated him and they'd say anything against him. You hear yarns about him making some gins sit up on top of a hot tin roof. I think that's all bull. They'd take getting up there for a start, and as fast as one got up the other buggers would be jumping off. You've only to work things out like that. He was supposed to have left them at the homestead to water the garden or some bloody thing, and when he came back he found they'd let it all die. This was over at Sturt Creek, somewhere about there.

> *Wason Byers, the part-owner of Coolibah Station on the Victoria River has been committed for trial on a charge of stealing 88 head of bullocks, the property of Bovril Estates ... It is alleged that the Coolibah brand MTQ, was burnt over the well-known Bullshead of V.R.D.*
>
> Hoofs and Horns, October 1953:67

Wason Byers was tried for duffing cattle once, when he and a bloke named Kenna were joint owners of Coolibah station. Wason wanted to make up a mob of bullocks to go to Dalmore Downs and they were sending cattle there from Coolibah. He was mustering cattle in there and just running them up in the bloody crush — cleanskins, VRD bullocks and everything.

My mate Roley Bowery was working for Wason at the time and he saw that things were getting too bloody hot, so he got to bloody hell out of it. Even the blacks knew some of the cattled belonged to VRD, but in those days they wouldn't say too much. Any VRD cattle, Wason would have the brand there and he was pretty cunning, he'd tell the blackfellas, "Go on, you brand'im". That way he wasn't doing the actual branding himself. They'd clamp the bloody hand brand on them to smudge out the VRD brand and then he'd cut off half the ear to get rid of the VRD earmark.

All the cattle had to be dipped at Anthony Lagoon, and of course when the brands are wet you can read them 100 yards off. Even if one brand has been placed over another, when they're wet you can see them both. Having half an ear missing looked suspicious too. Someone saw them along the road and recognised that some were from VRD, and of course, they wondered what was going on. One thing led to another and the police were called in.

Gordon Stott arrested Wason and held the cattle there, and eventually it went to court. Eight bloody days of it! Of course Wason turned round and said, "I wasn't there. I can't keep me eye on the blacks all the time. They must've branded 'em when I wasn't lookin'." Roley Bowery reckoned he saw one of these bloody hand brands over a VRD brand and it was plain as day, but Wason paid £600 a day for a solicitor from Adelaide to defend him, and he got out of it.

> *In his summing up the Judge said that he was not satisfied on the evidence of the natives alone that illegal branding of cattle took place at Coolibah or that Byers was present or had taken part in the branding.*
>
> Hoofs and Horns, December 1953:40–41

After the case was over and he was going along the road,

he's supposed to have pulled a bloody revolver out. They said, "What's that for?" and he said, "There's two bullets there. One was for Stott and the other was for meself if I'd got convicted."

A fella by the name of Alf Barwick was in Katherine one time. He was a small, slim-built, quiet sort of a bloke and a good ringer, but a bit on the surly side. Alf didn't care much for Wason Byers, and Wason didn't care that much for Alf. Wason was in Katherine at the same time as Alf one day, and he found out that Alf was coming out to VRD. He sought him out and said, "Hey Alf. I want you to do me a favour."

"Yeah, well all right."

"Look," he said, "it's to take a mare and foal out to Charlie Schultz. Will you drop them at Humbert for me?"

Alf had previously done a few days' stockwork on Humbert and he said, "I don't mind taking a mare, yes. Does it belong to Charlie?"

Wason replied, "No, but you drop it there. I'm going to give it to Charlie," or something like that. Wason told Alf that the mare was in the main horse paddock at Manbuloo Station, near Katherine, and he said, "Look, it's the only mare in there with a mule foal — a grey mare. You'll see it when you get there."

Alf said "I'll see the manager, I'll see old Tom Fisher about it."

"Oh no need to see Tom. I told him I was going to try to get yer to lift her. Just go through the paddock and pick it up. If it's gone he'll know that you got it."

Alf said, "Oh, that's okay then."

He went out there and got this bloody mare and foal, and brought them across to Humbert. He said, "Charlie, Wason Byers give me a mare and a mule foal to drop off with yer."

I said, "Who's it for?"

"Well, I thought it belonged to you?"

"Not to my knowledge."

"He must be giving it to yer then?"

"I don't want another bloody mule. I've got enough here as it is."

Alf said, "Wason told me to shove it up through the horse paddock and he'll be along later on."

"When's he due out?"

"I don't know."

"Are you gonna stop here and wait for him?"

He said, "No, I'm in a bit of a hurry so I'll go back to Gordon Creek now. I'm looking for a job and I'm hoping to get one at the Pigeon Hole or Montejinni."

As it turned out Barwick got the job at Montejinni. He didn't tell anyone where he was going, and he didn't know when he went over and joined Snowy Shaw that Wason Byers would be lifting cattle from Montejinni — taking them into Wyndham.

To Gordon Creek received information re grey mare and foal continued towards Humbert River ... Alf Barwick said the foal was given to him by Wason Byers in payment for breaking in horses.

Timber Creek Police Journal, 2nd and 4th December 1941

About three or four weeks later, Greg Ryall the policeman arrived at Humbert and he said, "Charlie, did Alf Barwick drop a mare and a mule foal here the other day when he passed through?"

"Yeah, it's here." I was looking at him half-stupid you know. "It's in the horse paddock — do you want it?"

"It belongs to Barwick doesn't it?"

"No, it belongs to Wason Byers."

"No," he said, "you're wrong. It belongs to Alf Barwick."

"Alf Barwick didn't tell me anything about owning it. He said he was dropping it here for Wason Byers."

"Oh did he? Well I'll be taking it."

I still didn't wake up. It wasn't till he was going that Greg said, "That mare and the mule foal are hot."

"What! Well, if they're hot they must have come from bloody Katherine."

"Yes, that mare belongs to Manbuloo."

"Look, I never even looked at the bloody brand. I just saw the horse running round there and didn't think anything more about it."

This bloody Wason Byers had tried to get Barwick hung! That's the sort of a bastard he was, you know. So anyway, Greg Ryall took the mare and foal down to the police station at Timber Creek, and later on he saw Wason about it. Of course, Byers denied he'd asked Alf to take the horse out and said, "That bloody Barwick, it was him that shook that bloody mare and now he's tryin' to put the blame onto me." He was a convincing liar too, the way he'd shake his head and rage and roar.

Of course, Barwick was questioned by the police too, and later on he struck Wason there at VRD. He went straight up to him and said, "You bloody mongrel bastard. You tried to bloody well get me hung didn't you!"

Wason Byers said, "You took that bloody mare your bloody self, you took it yourself," and when he went to walk away, Barwick walked around and stood in front of him. Everyone thought Byers would swing him one, but he didn't — he knew he was in the wrong of course.

Then Wason Byers turned up at Montejinni and Alf Barwick spotted him on the flat. He said to Snowy Shaw, "Is that that bloody Wason Byers over there?" Snowy told me later he bloody-near dropped in his tracks because Alf was telling him all the time, he'd say, "That bastard Wason Byers, a man should drill him, a man should drill him." And Snowy said, "Alf Barwick was a quiet sort of a bloke and just the sort of bastard that *would* pull a gun on someone."

Well, old Wason could be a jovial sort of a chap when he wanted. He came down to the stock camp and he was laughing: "Haw haw haw, g'day Snowy, g'day so-an-so, g'day Alf." Alf just picked up his cup of tea and ignored Wason. Snowy said later, he thought, "Ohh Jesus, this is going to be right on here. What am I to do?" He knew he couldn't stop Wason Byers because Wason was a great big man about fifteen stone. Anyway, Snowy sat down at the table and he reckoned Alf never opened his mouth. Wason would crack a joke and go, "Haw haw haw haw," but Alf Barwick would just glare and give a surly twitch to his shoulders. He'd look Wason Byers right in the eye and not a smile would crease his bloody face.

Snowy said to me, "Oh Charlie, look, if I'd had £1000 I'd have given it to be away somewhere else. I'd never been in a shoot-up before, but I was certain there was gonna be one there." According to Snowy, Wason Byers had the bloody wind up that Alf Barwick might bloody well shoot him too, but he got away with the cattle before anything happened. Yeah, that's the sort of bloke Byers was.

Mounted Constable Sheridan was a policeman based at Wave Hill when I first went out to Humbert River. He was in that country for years and I met him on different occasions. I thought we should respect policemen, but he certainly didn't respect others. For instance, in March of about 1935 I was on the boat coming back to Darwin after a holiday in Queensland, and Sheridan was also on board. Now Sheri was a terribly heavy drinker. When he got on the boat at Townsville he was full and he went on drinking from the time the boat left. He was as full as a goog the whole time and at breakfast he'd come in with only a singlet on, and this sort of thing.

Another time I was coming back from a holiday in Darwin. I used to arrange with my boys to have my horses sent in from Humbert to be at Katherine on a particular date. This time I came back earlier than planned, so my horses weren't waiting for me. Sheri was there when I arrived and he had a fella with him, a boring contractor who went by the name of Jerry O'Reily.

Jerry was a very nice chap but, ohh, he was a bugger on the grog too. He said to me, "How are you going out to Humbert, Charlie?"

I said, "Oh, the best way I can."

"What about your horses?"

"Well they're supposed to arrive next week."

"Care to throw your swag on the truck and come out with us?"

I said, "Right."

Sheri was going out to VRD because somebody had died, and he had to get there by a certain date to sell up the dead bloke's horses and camping gear. He was drunk from the time he left Katherine, and Jerry O'Reilly was pretty full too. Jerry had a big ten ton truck. He had a VRD stockman by the name of Jack Gallagher with him, a very dark-complexioned little bloke — a bloody good man at running a stock camp and they reckon he could ride like buggery. Jack had been in to Katherine for a booze up and was going back out to the Pigeon Hole. He was running the Pigeon Hole camp and before that he'd been running the Mistake Creek camp. There was Jack Gallagher, myself, Sheridan, and Tas Fitzer was on that bloody trip too! And, oh Christ, old Morris Bruten! Morri was a cook and one of the wittiest men I ever struck in the Territory. It didn't matter what Sheridan said, or Tas or any of them, Morri had an answer for it. He eventually died at the Timber Creek Depot and Tas Fitzer buried him near the old horse yard. He had guts — he knew he was dying and he was cracking jokes right to the last.

> *Left for V.R.D. in Jerry O'Reily lorry. The party was Sheridan & Tas Fitzer both policemen from Wave Hill Jerry O'Reily bore contractor, Jack Gallagher, Morris Bruten cook on Gordon Downs & self. camped King Crk.*
>
> Humbert River Station Diary, 27 September 1929

Well that was the team going out, and they bloody near drove me mad! I was the only one that didn't drink and they wanted to get me drunk, right or wrong. "Come on Schultzy," they'd say, but I just wouldn't even have the first one of course, let alone any more.

Left for Delemere but struck the road party 24 mile this side of Willeroo so that was as far as we got. Everybody drunk.

Humbert River Station Diary, 28 September 1929

Somewhere along the road Sheridan and Jerry O'Reily started mucking around, and Sheri bloody-well bit O'Reily's big toe. Oh, he was a mad bugger on booze! Then he bit Tas Fitzer on the side of the chest — you could see the mark of the full set of teeth. Sheri was one of those, when he was drunk he was laughing all the time, but he knew what he was talking about. Well he was supposed to get out to VRD on a certain date, but he was three days late getting there. It took us about six days to go from Katherine out to VRD. We went out on the Delamere road — those days there was no Dry River road — and by Christ, wasn't it rough!

They'd pull up round about eleven o'clock. "Dinner time," they'd say, and stop there all day, this sort of stunt you know. I was more or less the "Water Joe" — I was filling up the billycans and making tea for them and then I'd get to bloody hell out of it and sit under a bloody tree, and wait there until they got going again.

I'll admit I was frightened of what might happen, and I used to keep away by myself, but by Jesus I killed myself laughing at times over the things they were doing. Jerry would act the goat — he was just as big a wit as any of the others — and he'd say something, and somebody'd have an answer for him. And then they'd start rolling round, mucking round, and "What about another beer?" they'd say.

They'd say to me, "Hey. You had dinner yet?"

"No."

"Well get yourself something and get over there like a blackfella. You won't drink, you must be a blackfella. Keep away from us! Get under that bloody tree there and eat your tucker. We don't want you." And they'd be laughing, and the witty way they were saying things — by Christ Almighty! Then they'd come up later on, "Schultzy! We want a cup of tea. Haven't you got that bloody billycan boiled?" Well actually I almost always had the billy boiling whenever they came over for tea, and I was hopping in and getting my cut too — I had to or else I'd have bloody-well starved.

Stopped the day at Delemere Stn. everybody very drunk.

Humbert River Station Diary, 30 September 1929

Sheridan was a fighting man and when they got to Delamere they found a pair of boxing gloves. Jack Davidson was the manager of Delamere at the time, a big tall fella about six foot three, and he could fight too. Bugger me, Sheridan and Jack Davidson — both as full as boots — they had to put the gloves on. Oh Christ! I stood off about thirty yards while they bashed away at each other, and they were laughing. Sheridan would knock Jack Davidson down, and he'd laaaugh like buggery you know, and then they'd box away again. Next minute, down would go Sheridan, and then Davidson would sit back and laugh. Oh listen, I tell you I was that bloody frightened! There was blood flying everywhere, and then they'd patch each other up with sticking plaster and get stuck into it again.

Jerry O'Reily was stoomed. He lay down somewhere on the flat and there he bloody-well stopped! Then Morris Bruten said, "I don't think you could fight your way out of a paper bag Sheri." He had one or two drinks in him and he said, "Come on Sheri, I'll put the gloves on with yer." I thought, "Ohh, Holy Ghost! What's going to happen here?" That bloody Morri, he held his own with Sheri. By Jesus you know, I'd have liked to have seen him fight in his young days. He gave Sheridan all that Sheridan was giving him, and they were still laughing. Then Tas Fitzer had to put the gloves on. I think he put them on with Morri. Tas used to get pretty full, but he wouldn't flake out to it, he was one of that sort.

The part that struck me was that Sheridan was supposed to get out to VRD days before. I heard later that different ones turned up there at VRD and waited a day or two. Some wanted the horses and some wanted some of the packs, or something like that, but Sheri didn't turn up.

They woke up then that he was on the grog, especially when they realised he was getting a lift out with Jerry O'Reily. They thought that Tas might've been able to push things a bit, but Sheri was senior to Tas, so Tas had to more or less do what he was told although he was trying to pull things into gear. Now, if anyone else had not turned up on that particular day, Sheri would have reported them in Darwin. "Do as I say, not as I do," like the old parson, you see. That was Sheridan and he was a senior bloody policeman.

Christ Almighty! They got to bloody VRD and Jerry O'Reily's toe was that bloody bad where Sheri had bitten him that he had to go into hospital. They thought gangrene or tetanus was going to set in. He sobered up a bit there but he nearly drove the two sisters at the Inland Mission Hospital mad. He was a bloody hard case — look, it was really a hard case turnout, the whole lot of them! I was so pleased when we got to VRD and I found my horses there that I rode through to Humbert that night, and I was bloody pleased to see the last of them.

Not everything Sheridan did was crook of course. He did a bloody good job tracking down and saving Jackie Rogers who'd gone missing down south. Jackie was an only son and his parents spoilt him — anything young Jack did was always right.

They reckon Jackie was never that sane you know. The poor bugger went down to Adelaide one time and an uncle of his had him certified insane, and got him shoved in an asylum. Constable Sheridan was in Wave Hill at the time. He knew old man Rogers and he knew young Jack too. As a matter of fact he kicked him up the arse more than once when he was a kid running round the bloody pubs. After a while people realised that nobody had heard from Jackie for a good while, and everyone was asking Sheridan where Jackie Rogers had gotten to. One rumour had it that Jackie's uncle was after the station his father owned.

Sheri began making inquiries about what the hell had happened to him and it was just a complete blank. Jackie had been planning to down to Adelaide to see an auntie of his and Sheri thought, "I'll look into this." He was worried there might have been foul play. He got in touch with old Rogers' brother in Adelaide and told him: "I'll be in Adelaide at such and such a date." (Of course, you had to go by train from Alice Springs those days.) "I want you to meet me at the railway station."

So this bloke met Sheridan and Sheri asked, "Do you know where Jackie Rogers is?"

"Well, listen Mr Sheridan, he was putting on a hell of a turn here and …"

"Where is he?"

"I was gonna get onto that. He was putting on a hell of a turn here and I got him certified insane. He was dangerous!"

Sheri said, *"What did you do?"*

"I got him certified."

"Where is he now?"

"In the asylum."

Sheri said, "Right. We're going straight out there now."

"But it's night time!"

"I don't care a bugger. We're going straight out there now."

They got a taxi and went straight down to the asylum. When they got there Sheri said: "I'm a senior constable from the Northern Territory and I want to see Jackie Rogers. Where is he?" The lady in charge said, "He's out the back there somewhere," and she called Jackie out.

Jackie came through the door and looked, and they reckon he just went bloody mad when he saw Sheridan. He raced straight up and threw his arms around him and said, "Sheri, Sheri, get me out of here! Get me out of here." Then he looked at his bloody uncle. Ohh, Jesus Christ! He said, "That bloody mongrel there called me insane."

"What's the matter with yer?" Sheri asked him.

"These bastards, they've got me here, and I don't know

when they're gonna let me out, but they tell me if I work hard enough I can go home."

I believe the asylum people reckoned Jackie was the best worker under the sun. No doubt he was a bit simple, Jack, because they told him they'd let him go if he worked hard enough, and he believed them. He'd been in there about three bloody months then and still working.

Sheri turned round and said to the woman in charge of the asylum, "I'm taking this bloke home. He's my responsibility from now on and I don't care what you think."

"But — he's insane!"

Sheridan said, "There's a lot of us like that in the Northern Territory."

Bloody Sheridan was a big red-faced wild Irishman you know, and they reckon those people in Adelaide talked that over for years and years. "This bloody policeman came down from the Northern Territory and took a madman out of the asylum, and he reckoned he was insane himself! 'There's a bloody lot of us!' he said, this mad policeman."

After Jackie got out of the asylum they gave him a certificate to say he was quite okay to leave. Otherwise he could've been arrested again on the outside. For years afterwards, if Jackie ever got into an argument with anyone he'd turn round and say to them: "I'm a bit more saner than you. At least I've got a certificate to say I'm sane," and he'd pull it out and show them.

Later on Jackie Rogers was the head stockman at Moolooloo and his horse turned turtle with him there and killed him. One side of his head was caved in. They brought him into Katherine by Flying Doctor and he lived another nine days. When they held the post-mortem the coroner remarked that Jackie must have hit a very round stone, because he had a round depression on the side of his head. The basalt stones in that country are all round.

Jackie was a good ringer, but he was always raging and roaring at the bloody blacks, that sort of turnout. They reckon if he got mad at them he'd pull his revolver out.

Well, the whites out there put their heads together and to this day they agree that he got a buster all right, but they reckon a bloody blackfella finished him with a stone.

Poor old Burt Drew was cooking at Moolooloo at the time and he took Jackie's death very hard. He was deaf as an old crow you know — he was hopeless! — and they reckon he'd wake up during the night and sing out, "Poor old Jackie Rogers." Being deaf he wouldn't know he was singing out so loud, and they reckon it was like a bloody ghost or something. He'd be there and, "Poor old Jack Rogers," he'd say, "Poor old Jack Rogers."

13 They Thought I'd Gone Mad

> *Mr. Schultz's life ambition is to have the best cattle station in the Northern Territory.*
>
> The Northern Territory News, 4 December 1952

I was the first to bring cattle and horses into the Northern Territory by truck. After I got married I got a four-ton Lend-Lease Chevrolet truck from a bloke in Mount Isa. It was really a necessity, but I was a long time without spare cash and I didn't know whether I was spending too much money. At last I decided to get it and I paid the big tally of 380 quid for it. That was back in the war years, so I was lucky to get it. I thought I was made, and I was too because I could do things so much easier then.

Around 1947 or '48 I brought an Illawarra bull and three heifers.[1] They came by rail from north Queensland to Mount Isa, and when I put them onto my truck the eyes of the locals stuck out like crabs. I think they thought I'd gone mad! Those days, of course, with the bad roads it took four to five days to make the trip one way. Later on we used to get quite a few out on the truck at one time. On one trip I picked up fourteen Devon bulls in Mount Isa.

> *Charlie Schultz of Humbert River was an interested spectator at the Annual Yearling Sales held in Adelaide.*
>
> Hoofs and Horns, April 1959:59–60

Over a period of years I brought six or seven stallions and a number of mares out by truck. I got four stallions from Queensland and about three from South Australia. I'd get the stallions railed in to either Mount Isa or Alice Springs and pick them up there. Some of the mares I trucked all the way from Adelaide. I really had no other way of getting a horse up from South — it was too far to walk or ride them. Where there was a yard I'd unload the horses at night, but odd places I let them stop on the truck all night. When I was first trucking horses from Adelaide to Humbert River, I used to try to take them off every night until on one trip Donna said to me, "Dad, when you went to India, how long were those horses standing side by side in the hold?" I told her they were on their feet for the whole nineteen days. Donna pointed out to me, "Yet between Adelaide and Humbert you take the horses off every night?"

That gave me food for thought. From then on I'd run straight through to Alice Springs before I'd take them off. I'd take them off there for the night and the stock inspector would have a look at them the next day. Then I'd go from Alice Springs right through to Humbert.

Going in to Mount Isa one time to pick up one of those stallions, I nearly got myself hung! I got in late, around about four, and as I was passing the trucking yards I saw a stallion there. I thought, "Hello, that'll be mine." I went over and had a look at him, and I thought, "I wonder how long he's been here because some good-natured bloke's been feeding him." I could see that someone had given him some chaff and oats. I thought it must have been one of the railway blokes doing that off his own bat, and I reckoned I'd find out who it was and what I owed him you see. I booked in to a hotel and when I came back later that evening to give him a night's feed, I saw that he'd been fed again.

When I got there a Wave Hill bloke called Paddy Conway was just about to take off. I said, "What are you doing?"

"Just feeding the stallion."

"Oh, thanks Paddy. I'd been wondering whether he'd

been fed or not. What do I owe you for it?" I could hear him doing some heavy thinking — I think he thought I'd gone mad — and he said, "What do you mean? What made you think I didn't want to feed him?"

"Well, why should you be worrying about my horse?"

"I think you're mistaken. This is a Wave Hill horse and I'm taking him back in the morning."

"Jesus Christ! That's my bloody horse!"

"No, it's a Wave Hill horse."

"Like bloody hell," I said, and I produced the weigh-bill.

Well, we eventually got the situation sorted out. Wave Hill had a stallion coming out too, but my horse had been unloaded at Cloncurry by mistake. I'd been up to the Railway Department office earlier and I got the weigh-bill for my horse, but they made a mistake and told me the horse in the yard was mine. Christ, if I'd been going out to Humbert that evening I'd have put him straight onto the truck and left town. There'd have been a hell of a stink! I can just imagine getting out near Alexandria Station or somewhere, and a policeman coming along and saying, "Hey! Come here. What've you got there?"

"A stallion."

"What are you doing with him?"

"Taking him back to Humbert."

"No you're not. What are you doing trying to lift a Wave Hill stallion?"

I had the weigh-bill for my horse, but of course, a lot of people always blamed me for stealing that Wave Hill stallion years before, so everyone would have reckoned I'd gotten a bit cheeky and tried to do it again. A bloke would have had to be damned cheeky all right! — and bloody stupid as well.

... not content with the use of the universally-used bloodwood posts, Chas. has gone one better and built his yards of bloodwood

posts and Lancewood rails, carting all his rails from No. 13 bore on the Murranji.

Hoofs and Horns, May 1951:46

It was after I got the Lend-Lease truck that I decided to build a yard with lancewood rails. Ironwood was one of the best yard timbers, but there's no ironwood on VRD and none around Humbert either. There's ironwood on the Murranji of course — it was always a problem there for drovers because it's poisonous. I always remember a bloke who came over from the west. He had five camels with him, but by the time he got through to the other side of the Murranji Waterhole he only had two left. The others got stuck into the ironwood and it killed them.

They had a lot of coolibah swamps on VRD, so they used coolibah posts and rails, and by Jesus they're solid. White ants don't like them either. There's no coolibah around Humbert at all and the best local timber for yard rails was bloodwood, but it doesn't last long, and anyway the best timber was all cut around there. I was sick and tired of putting new bloodwood rails in my yards every second or third year, so I decided to build a yard with lancewood rails. Lancewood has a lifespan of about twenty-five years and the white ants don't seem to attack it the way they do bloodwood. The nearest lancewood was at Number 13 bore on the Murranji Track. That's 150 miles from Humbert. I had it in the back of my mind that if I could get 100 rails at a time it would be worthwhile. When I got the Lend-Lease truck I decided to give it a go.

Well I had to find it all out by trial and error — how many I could get on the truck and so on. I went over and discovered I could carry 100 rails. For each load, the first day we'd leave Humbert and go as far as Moolooloo — it was an all-day sitting getting that far. We camped there and the next day we'd get into Number 13 bore around five o'clock. Most of that country was Crown Land then so we didn't have to ask a landowner's permission to cut the timber. Somebody would put on the beef while the rest of us started cutting rails. I had three boys and I had to show them which trees to cut because they were inclined to cut them too big — that wasn't necessary with lancewood rails. Or else they'd be cutting them too small. I'd go round and say, "Right-oh, this one here," and then I'd go onto the next one. In the end I woke up and I'd use an axe to mark the trees for them to cut down.

Usually we worked it that we'd have the rails loaded and get back to Moolooloo again that night. That was a hell of a big day you know. There were no bulldozers or graders those days, and by Christ some of those creeks were bad between the Murranji and the Humbert. There was a bad one at the Murranji jump-up, and at odd times I got caught at Waterbag Creek, especially if there'd been rain about. I got stuck a few times on the banks at Dashwood Crossing. Peartree Creek between VRD and Dashwood Crossing wasn't too bad, and then I'd go right through to Steep Creek, down past the Gordon Creek homestead.

Steep Creek was the worst of the lot. I'd have to throw off about fifty rails there. I'd get down to the bottom and take off up the other side, but time and again I'd only get halfway up. Generally there were four of us, three boys and myself, and we'd have to carry each rail about thirty yards up the bank through dust about a foot deep. We'd load it on the truck again and then walk back for another one.

I was the first to bring lancewood rails into that country. About three or four years later, after they saw that I could do it, some of the other stations started going to the Murranji for rails. They put up a yard with lancewood rails at VRD head station. I could never make out why they hadn't done it before because they had good strong mule teams and later on they had really big trucks.

I carted the rails in October or November, and over the wet I'd keep two boys back with me while the rest went on walkabout. I'd have about 400 rails dumped on the ground at the yard and we'd get stuck into it. I'd be up at the stockyard adzing rails down and putting them in. They were all big rails and very heavy, and it was so stinking hot I'd just poke along, taking my time. If I could get in about ten or fifteen in the morning I reckoned that was good going. That was about three panels. Then of an afternoon I'd go up again and put in another two panels. That'd be about twenty-five or thirty I'd put in — if I got twenty-five rails in I was very happy.

We'd work till about twelve o'clock and then knock off till it cooled down a bit, around about three o'clock. Then we'd work till dark. After tea we'd go back to the yard and so that we could see what we were doing we'd light big fires. We'd hop into it again and do another couple of hours. That would make up for the long afternoon break and at least it was nice and cool — everyone liked working after tea.

The drafting yard on the Humbert has long been recognised as one of the best planned yards in the country and so impressed Colonel Rose when head of the Animal Industry Division that he had plans of the yard blue-printed and distributed as examples of how a drafting yard should be laid out.

Hoofs and Horns, April 1960:62

Another thing I used the Lend-Lease truck for was to power a saw. We had a saw bench at the station that my father and I had brought out from Queensland, but we had no engine to power it. When we got the truck we took the tyre off one of the hubs, jacked the back wheel up and ran the saw off that. We had all the power under the sun then and by gee we did a lot of work. We burnt a lot of petrol too, of course.

In the wet season we'd go out several miles, cut down twelve to fifteen big carbean trees, roll them up onto the truck and bring them home. Then we'd work for about a fortnight or so, cutting the logs into big slabs ten to twelve inches by two inches and about eleven feet long. As they came off the saw bench I had men making gates for yards. Sawn timber of this size made a good heavy gate that would stand up to the weight of any bull. We painted them with either sump oil or coal tar to protect them from the weather.

> *When we got to Victoria Downs Station [in 1909] ... we found that the manager had installed a small steam engine, and with the help of a circular saw and a number of blackboys was cutting planks out of paper bark tree logs.*
>
> L. Gee, 1926:25

Years ago VRD cut paperbark trees on Humbert station near the crossing at Williams yard. It's spring country on the Wickham River and the best paperbarks grow there. Oh, they were beautiful sticks, thirty or forty feet long and straight as a gun barrel. They had plenty of labour and they sawed them down with a crosscut saw. Then they got mule teams hooked onto the logs and snigged them out and rolled them up onto a wagon. When they kicked off for VRD it would take them more than two days to get home.

There was a saw bench at VRD and they rolled the logs off there. They sawed them up and had all the timber they wanted. They cut timber for two homesteads, one for VRD and the other for Pigeon Hole. Pigeon Hole was one of their main camps then — both homesteads were two storeys and exactly the same design.

> *Last month the Territory had its first cattle air-lift ... The cattle concerned came from Fossil downs, in the West Kimberleys, across to Humbert River. They were landed on the V.R.D. drome by Bristol Freighter, and were dark red polled shorthorns for Charlie Schultz.*
>
> Hoofs and Horns, August 1951:43

Round about 1950 I decided to buy a good number of bulls to put some fresh blood into my herd. To get them onto Humbert I would have had to walk them out from Queensland, but that took too much time. Of course if you want to get a thing done you just damn well do it, and that's all there is to it, but I was wondering for a long time if I could do it any other way. I knew I could put two or three on a truck — but I wanted a lot more than that.

Then I heard about beef carcases being airlifted from Glenroy in Western Australia into Wyndham. They were killing cattle on Fitzroy or Mount House and flying twelve or fifteen tons into Wyndham. I knew there was a cattle stud on Fossil Downs and I thought, "By gee, that means that if I got very young bulls I could get a lot on" — I could sacrifice age to get quantity, you see. I'd only have to wait twelve months and they'd be old enough to work on cows.

So I wrote over to a fella by the name of Ian Grabowsky, or "Grab" as they called him, a great airman in his day.[2] Grab gave me the particulars of the costs involved and it looked like it'd be worth flying cattle over, considering the time and money it would cost to drove them. Grab wanted to boost the airlifting business, and as soon as I wrote and mentioned that I was thinking about getting bulls over from Fossil Downs, he couldn't get to me quick enough!

"How many do you want to get over? And what age?" he asked me.

I said, "What about two year olds?" and told him the quantity.

Bill McDonald was the manager-owner of Fossil Downs. His forebears were the first ones that brought the cattle into the Ord River country. He wanted to boost his stud sales over there so he was heavily involved too. Bill and his wife spoke to me on the radio about getting the bulls over. I said, "How are we going to do it? You can't very well let them run riot in the plane."

"Well what about some little yards, and we'll tie them in four abreast?" He reckoned they wouldn't be able to do any harm that way.

I said, "It could be all right in a pinch, but what about breaking the buggers in to a halter lead?"

"Well, that'll be a good idea."

"Look, if you've got the halters, break them in."

Cattle are just like a damn horse — to break them in to a halter you put the halter on them, tie the buggers up and go away from them. You don't tease them or annoy them, because if you do they start fighting and pulling back and bellowing. Once they find they're tied up they quieten down, so about six weeks before they were put on the plane, Bill got a New Zealand lass to start breaking them in to halters.[3]

When everything was ready to go, Grab organised to get a plane on a weekend when they weren't being used to fly beef carcasses. He had the meat-hooks taken out of the plane on a Friday and then he and Bill McDonald made a framework of five breast-bars that went across the plane.

Four or five bulls were halter-tied to each plank and they just got in twenty neatly, but they had to take two out because there was too much weight on the plane. Bill said, "You'll have to take younger bulls."

"I'd like to have got them around two years old," I said, "but it makes no difference to me." I just wanted the bulls on Humbert. The nearest airstrip those days was at VRD. He said, "Have they got a crush to unload them?" I said, "No, but I'll just run them down a plank — they shouldn't be any trouble." Even if they did fall off, it was only about a two-foot drop.

It took four and a half hours to get the first load of bulls to VRD, and they never shifted coming across. In fact they reckon one or two of them even lay down. Coming over, the freighter had to divert to Argyle Station to refuel, but I think on the way back it flew direct because it had no load on going that way. They got in round about twelve o'clock and of course the VRD mob came over to see them unloaded — they were a bit jealous you know. Magnussen was the manager there at the time and he came over and said, "Charlie, about the pilot and crew — I presume you'll be looking after them?" He was worried I might impose on VRD for lunch for the aircrew. I said, "It's all right Mr Magnussen, we've got everything in the car there. The car will even take them down to the river, and if you don't mind they might even swim in your waterholes." He looked at me then, the bastard, but he never answered.

When they got to VRD I had all my boys there and we ran the bulls off in fours, down the little ramp that had been used to run them up into the plane. The first four took fright and pulled a bit, but with three or four boys hanging onto the end of the halter-rope they steadied down. I waited until the next four came off, then said, "Right-oh, now we'll slip the halters off and take them away quietly." They didn't realise the halters were off and they kept together. We got them all off and they just walked away as though they'd been driven there for years.

We had lunch there and then Bigfoot and Toby began to shift the cattle over to Humbert. I told them, "Go to the Gordon Creek Camp tonight and tomorrow go to Flying Fox yard". (That's a yard five miles on the VRD side of Humbert.) "Yard them there and next day take them on to the station." I kept them there for about four or five days just to get them used to the place.

Eventually I took that first twenty out to Ivnors Pocket and let them go. There's usually a big mob of cattle on a swamp there near Flying Fox yard. I said to the boys, "Now, when those bush cattle start to run away, just round them up and hold them there for a while." They let the bulls go amongst the old cattle and they just spread out and stopped there.

That was the first fresh blood put into the Humbert River cattle, and they were the first cattle to be airlifted into the Northern Territory. About three weeks later the Bristol Freighter was ready again and I got another load over.

The last to come over was a mob of fourteen cows. They were a lot bigger than the bulls. By Christ they were big heifers! They had a different pilot that time and he got bushed. They were late coming — twelve o'clock came, one o'clock, and still they didn't turn up. We thought, "Holy Ghost, what's happened?" We knew they were coming because we'd arranged to speak with Fossil Downs on the radio before the plane left. They'd told us,

"They're all loaded and they're taking off. They should be there by about half-past eleven."

When twelve o'clock came they were flying round near the bloody Pigeon Hole. For some reason they missed VRD and ended up at Timber Creek. They didn't know where they were, so they landed to find out. The police came out to the Timber Creek airstrip and the pilot said, "Your airstrip's very short!"

"Yes, it's less than 800 yards."

"Well! We have to bloody well get out of this now."

By Jesus, they only just got out of that! They took the plane right back off the end of the strip and rushed it. The Timber Creek airstrip is in amongst mountains — they reckon the plane only cleared the ranges by about twenty feet. The aircrew said the bloody cows were sitting back on their backsides and you could hear them slipping and sliding. They landed at VRD around about half-past two, unloaded them straight away and let them go. Then they said, "We're taking off again straight away. We have to get back before dark."

A year or two later they held a bull sale at Brunette Downs and a lot of the studs flew cattle up from Victoria[4]. They started to brag a bit then, claiming they were the first to airlift cattle to the Northern Territory, but old Colonel Rose put his spokes in and said, "Oh no. Humbert River was the first to airlift cattle."[5]

Some years after that airlift from Fossil Downs, I pioneered the lifting of a big mob out of the Victoria River district by truck — 560 or 600 bullocks. I remember when Johannsen in Alice Springs got the first road trains going.[6] I saw photos of them and I just shook my head. I thought they were a wonderful thing, but different blokes would say, "Ohh, they're no good. Your cattle get all bruised." There was a certain amount of bruising but not to any great extent.

They were experimenting with road trains all the time

around Alice Springs and gradually they got them going. Of course, the roads weren't as bad in the centre as they were up north where you had big rivers and hundreds of creeks and gullies. Eventually I decided to give them a go. At that time we couldn't get big trucks through to Humbert — only as far as Steep Creek — and the VRD manager then, Jack Quirk, he wouldn't let me use any of the VRD yards. We'd had that falling out over one of my Aboriginal stockmen, the incident that led me to build my own airstrip, and we still weren't what you'd call real friendly.

Well, we wanted to put up a loading ramp there at the yards at Gordon Creek. This would have been very handy for Jack later on, but he said, "No, I won't hear of it, won't hear of it!"

I said, "Can we build a bit of a wire yard the VRD side of Steep Creek?" — That's still on VRD country.

He said, "What are you going to do about a loading ramp?"

"We can make one with antbed I suppose."

"Yes, you can do that," he said. He thought we wouldn't bloody well have a crack at it. Well, I brought three coils of wire and some sawn-timber gates down from Humbert, and put up a small yard.

McArdle and Sullivan were the buyers, and I admired them because they had the guts to come into the VRD country and get those cattle. They sent twelve small trucks out to lift the bullocks — they could only put on about sixteen or eighteen head, so we shuttle-serviced them. The first lot went as far as Number 7 bore on the stock route east of Newcastle Waters, where there were some beautiful big steel yards. There they were unloaded and dipped while the truck came back to Humbert for another load.

When they got to the jump-up on the Murranji Track the trucks couldn't get up, so they got one behind the other and pushed the leading truck up. Then the leading truck hooked a rope onto the trucks at the bottom and pulled them up. When each truck reached the top it went on, so as not to waste time.

There were no bulldozers or graders then and no made-up roads, and by jove the creeks were bad. Between Humbert and Newcastle Waters the trucks chopped up the road in places — eventually parts of the road were churned into dust about eighteen inches deep. I lost very heavily there with bullocks smothered in the dust. Twenty-seven head of big fat bullocks I lost — a terrible thing but still, it was an experience. Once they got to Number 7, or even to Newcastle Waters, I don't think another beast was lost.

They had to go through to the Camooweal where they were unloaded and dipped again. Somebody would tail them out on the flat to let them have a feed and then they trucked them through to Stanford, near Cloncurry. The cattle were kept there for a fortnight or three weeks to brighten up before they continued down to the Ross River Meatworks in Townsville. The buyers didn't make that much money on them, but they certainly cleared expenses. Of course, Jack Quirk reckoned we were buggering up the roads and shouldn't be allowed on them. A strange man, old Jack, hard to get on with. He was all right as long as everything was going his way. You see quite a few of that sort of bloke.

Until this lot of cattle went out there were a few mobs being lifted by Johannsen in the Alice Springs district, but he didn't like to go out that far. He was striking the same trouble that I had with bad roads, you see. Once they saw that cattle could be shifted by truck up in the Victoria River district they really started to upgrade the roads. When they got the Katherine meatworks going the government really got a move on, and when the bitumen went through stations were sending cattle right across from Western Australia.[7] It's wonderful what good roads will do for you.

14 Bullita and Whitewater

Connor, Doherty and Durack were the first to take up Bullita in the early days, and in the 1940s they abandoned it. I was in the Lands Department in Darwin one day and they said to me, "Charlie, why don't you take up Bullita?" Actually I wasn't keen on taking it up because most of it wasn't very good country — in fact it used to be known as "The Rock". I wanted a better block, but I reckoned that if I got Bullita the Lands Department would think I had enough land, and there wouldn't be much of a chance to get anything better. Well there was nothing else going and I needed extra country, so I took it up.[1]

A few years later some blocks of top quality VRD land were being resumed.[2] These were the Camfield, Montejinni and Killarney blocks, and I was very keen to get one of them. When they had the ballots for the excisions there were thirty-two in for Camfield. Eventually the applicants were cut down to a chap by the name of Jim Edwards, and myself. For a while they were undecided who to give it to, but they ended up giving it to Edwards. A mate of his, Charlie Campbell, was supposed to be a shareholder too.

It was a hell of a blow to me because right to the last minute I thought I was going to pull it off. After I heard everyone else's evidence, I could see they didn't have the know-how for a start, and they didn't have the stock. I was close by on Humbert there, and I could have shifted my stock straight over. Halfway through the court proceedings Edwards had to catch a plane back to Queensland, and he came round and congratulated me. He said, "Well I heard your evidence and I'm gone. You've got it all right and I'm satisfied about that" — or words to that effect. Well there's many a slip between the cup and the lip.

> *VRD, with such a large herd of cattle, were on the whole, hopelessly out of hand, there being 90,000 on the books — others estimated 120,000 — which duly wandered everywhere.*
>
> F.H. Bauer, Report on Humbert River Station, 1957

Getting those blocks was the best thing that could ever have happened to the blokes who drew them. When VRD fenced off Killarney, Montejinni and Camfield, the new owners were able to just run their fences straight off the VRD boundary. Then all they had to do was put a few more fences up inside their own blocks.

Nobody ever knew how many cattle were on VRD. The old-timers reckoned there were 90,000 or 94,000 on the books, but since they put the helicopters into them they reckon there's anything up to 120,000 cattle. By Jesus, the bullocks that used to turn up on that place you know, year after year. Even in places where they weren't supposed to be. Some were chasing ten years old, with great long horns on them. By God, I tell you what, they had some beautiful bullocks.

VRD had a certain time to muster the blocks and if they weren't finished when the time was up, well that was just too bad for VRD. They knew it wasn't worth doing anything until they got a boundary fence put up so that went in first. Then they mustered and put the cattle on the VRD side of the boundary. Of course, a lot of the cattle they were getting were cleanskins.

The VRD mob had mustered and rounded up a mob of cattle and were still holding them on the Camfield block when the deadline passed. Jim Edwards, Charlie Campbell and I think Bill McDonald were there with several other stockmen. They galloped into the cattle singing out and cracking their whips and firing a few shots from their revolvers too, and they frightened the cattle hell, west and crooked. Of course, the VRD blokes weren't too bloody happy about lead flying around. They were inclined to just let them go.

After that big dust-up VRD went a lot quieter with Killarney and Montejinni. Each block was supposed to be cleared of cattle, but on Montejinni they did a muster and branded over 5000 head, I think it was. It was the same on Killarney — they branded near enough to 4500 and on Camfield about 5500.

It'll give you an idea of the cattle that were there because VRD had good brandings, yet Montejinni, Killarney and Camfield branded over 14,000 head of cattle between them before they brought any of their own stock onto the place.

It was 1952, the year of the big drought. An area noted for its birdlife was without a bird of any description.

The Northern Territory News, 30 August 1974

The cyclone rains again missed most of the Barkly and Victoria River districts. Trucks have been transporting feed to save £50,000 worth of herd bulls on Wave Hill ...

Hoofs and Horns, June 1954:38

Well, I missed out on the Camfield block I was after so I carried on with what I already had. Bullita was mostly very rough country, but at least it was well-watered, and there was one year when that water came in very handy. You could always count on a drought about every fifth or sixth year. In about 1953 or '54 VRD struck it pretty dry, but not as bad as just around the station there at the Humbert. It completely missed the rain that year and was totally bare. Bullita was always very lucky with storms, possibly because of the ranges.

That time I shifted over 6000 head from Humbert in lots of 500. I took them right up the Top Humbert and pushed them over a bit of a jump-up onto Spring Creek. When they went down they stopped down — they were that darn footsore because of the stones they had to travel over, they were only too pleased to stop! There was plenty of old grass and plenty of water in Spring Creek itself, and they went up all the valleys there and got all the feed they wanted. It wasn't what you'd call good feed, but at least it stopped them from being hungry. No doubt I saved a hell of a lot of cattle.

It's an amazing thing, I had to shift all the cattle out of my bullock paddock because the water went dry there. I shifted all my breeders first — they were the ones I was really looking after — and the last mob I shifted over was round about 800 bullocks and steers. I let them go in the Bullita bullock paddock, which was long and narrow. I was a bit dubious that there'd be enough feed for so many. I watched them closely there and saw they got a lot of top feed — bushes and trees.

I knew a certain amount would come back once the storms fell, but when I mustered to take them back to Humbert, out of the 800 I only picked up about 200. I thought, "Oh the other 600 got out" — right at the very back there's a place where they can get up on top of the range. I think most of them worked their way back to the Humbert bullock paddock during the wet, because afterwards I found about 700 or 800 head of male cattle there when I mustered.

It was the same story all over Bullita. Of the 6000 head of cattle we'd taken over I don't think we got any more than about 700 or 800 back. When good storm rains fell the rest gradually worked their way back from Bullita to Humbert, a distance of forty or fifty miles. It wasn't the distance that was surprising so much as the rough ranges they had to get through to make it back to their old beats.

Charlie Schultz, owner of Humbert River Station, had the bad luck to be horned in the leg by a bullock, but first aid was forthcoming when Mrs McColl, an ex-nurse, made the trip over to Humbert from V.R.D.

Hoofs and Horns, August 1953:17

One of the most outstanding experiences I had in forty-four years of station life happened when I was making up a mob for Wyndham one time. I had a knocked up plant — a couple of lubras, Betty and Donna, a couple of half-caste lads that Hessie was rearing, and Hessie herself in the mustering camp, cooking. We'd nearly completed our muster — we already had about 700 in the main mob which was in the bullock paddock — but needed to pick up another forty head. We mustered Ivnors Pocket and got about twenty-five bullocks and ten or fifteen spayed cows, and put them in Flying Fox yard for the night.

In the morning when we were letting them out, a young bull about sixteen months old, a little stumpy-horned bull, broke from the mob. I lapped him right around and was bringing him back to the mob when he whipped straight around and came onto me. I could see that he was going to hit my mare and I pulled the reins, but my horse was one of those swivel-necked mares — you could pull her head around but her body wouldn't come with her. I came straight onto him and there was a loud sharp *bang*. I knew straight away that I was hit hard. I forget whether I let him go or what I did, but I sang out to my boys: "Hold them here and let them settle down." They were pretty lively you see, and if you hold them in a mob for a while they become much quieter.

I started to ride in to the camp and I could feel blood trickling into my elastic-side boot, but I wasn't looking down at my leg. Betty was there when I was hit and she took off to the mustering camp ahead of me and called out, "Mummy, Daddy's been horned by a bull!" The pain caught up with me when I was about fifty yards off the camp — I was doubled up and nearly falling off my horse, and Hessie thought I'd got it in the guts. She and a couple of lubras ran over to me and of course my trousers were covered with blood. She said, "What happened?" I said, "Got horned in me leg by a bloody bull."

I got to the camp and slipped off my horse. I thought I'd have been able to hobble across to the tent we'd rigged, but no way! I said, "Would you get me to that bed in there." There were a couple of gins in the camp and between them and Hessie they got me to the bed. I laid down and then Hessie said, "I've got to take your trousers off." She began pulling them down and all of a sudden I felt my toes curling. I was in a lot of pain and I said, "Christ! What are yer doin'?" She said, "There's a piece of your trousers inside the wound, way in, and the blood has clotted around it." When she pulled the cloth it must have been pulling a nerve because it was making my toes curl up.

The wound itself was a round hole about two inches across — you never saw anything like it in all your life! When I had a look at it I thought, "Oh God, this'll never glue up. I'll always have a hole in me leg!"

Hessie said, "We'll have to take you in to the station." We made it in and then she got on the radio. She found out that the Flying Doctor had left Auvergne Station for Darwin fifteen minutes before she called, and she wasn't able to make contact with him. Instead, she contacted Fred Ryle at the Royal Flying Doctor Service in Wyndham.[3] She said, "Charlie's been horned by a bull and he's worried because he has to take about 700 bullocks to Wyndham. There's no one else to do it."

"Have you got any penicilin?"

"No. We're just bathing it in hot water."

"Well he needs some penicillin."

She switched off and straight away there were about three calls from all over the country from stations that had penicillin, but the trouble was to get it to Humbert. Auvergne called, but they'd have had to come the best part of 140 miles on the terrible roads of those days. Inverway called, but they were even further away. Scott McColl called from VRD and said, "My wife's a trained sister and we've got penicillin here. We're the closest, so we'll whip over straight away.

VRD was only thirty-five miles from Humbert. They turned up about an hour and a half later and gave me a needle. Then they showed my wife how to do it, but she said, "Oh, I don't know how I'm going to get on, I can't give anyone needles."

I said, "Look! Jab the bloody thing in me arm, that's all there is to it."

"Oh I can't."

"Listen, any other time you'd be hitting me over the bloody head with a stick or something!" I was trying to crack a joke you know, but anyway I think one of the lubras ended up giving me the shot. I had to have a number of shots and after each one I felt damned crook. Then I started to break out in hives and blotches from the effects of them, so Hessie got back to Fred Ryle.

He said, "Do you know if Charlie's allergic to penicillin?"

"No, I don't know."

"Well, it looks as though he must be."

As soon as I stopped getting the shots I came good.

A couple of days after I was gored the bullocks were ready to go into Wyndham. It was a twenty-eight day trip in and I had no one else to take them, so I bandaged the wound up and we started off. When I wanted to mount my horse I'd have to get on a log and tell one of the boys to come over. I'd jump and he'd grab my good leg and throw it over the saddle. Once I was mounted I couldn't feel any pain in the leg, but I soon discovered I couldn't bandage it properly. The bloody thing kept coming off all the time. I'd be riding along and the next minute I'd look down and find it hanging down a couple of feet.

When you're droving like that there's no spare time, you can't always be having a bath or anything. The best I could do was bathe the wound in hot water of a night time when I got into camp. By gee, I tell you what, riding a sweaty horse and with the bandage hanging down, I was really frightened of getting tetanus. Oh Jesus, I rode all the way into bloody Wyndham like that! And because we were short-handed I had no option but to do a night watch as well, along with the others.

We eventually got into Wyndham, and was I pleased to see the last of those cattle! I still couldn't walk, so they took me down to the slaughter yard in the Land Rover and I saw the last of them there. I was damn pleased when I finally got back home. It was the best part of two to three months before I could really walk. I always thought if you got a horn through a muscle, especially the leg, you'd still be able to walk. If it'd been any one else I'd have said, "Oh, I bet I could have walked on it," but by gee, I wasn't in the hunt. No way!

For years everyone knew that after Jim Crisp was speared he'd been buried up the East Baines River, but nobody seemed to know just where. By the time I got to Bullita most of the old hands had died or left the district. There were only one or two whites that ever really knew where he was buried in the first place.

At different times at Humbert or on the East Baines, I'd ask my stockboys, "Where Jim Crisp dead?"

"Oh, him buried, him grave that way somewhere."

No doubt the Bullita boys told them where it was and no doubt they'd mustered past it, but I think that because Crisp was speared by the blacks my boys didn't like telling me about it. It was only sheer luck that I found it.

After I acquired Bullita I mustered that country for years and never came across it. Then one evening, about 1950, I was mustering some cattle along the East Baines River when something unusual caught my eye. I looked and I thought, "That post has been cut with an axe! What's it standing there for?" I had another look and it suddenly dawned on me: "That's Jim Crisp's grave!"

I rode over, tied my horse up and had a good look. One post was still standing and I could see the remains of three other posts burnt off practically to ground level. The rails had all fallen down and there was long grass growing through it. The river was about seventy or eighty yards away and I thought, "In time to come this grave is going to get lost." There are big flagstones in the river there and I carried out three of the biggest I could handle. I made three separate trips down to the river and I put the stones on Crisp's grave, you see. I always reckoned that some day I'd put a tombstone or some sort of a marker there too. Well, eventually I did get a plaque done when I was down in Adelaide one time. I just put his name and the date he was speared by blacks, and my initials, C.N.S., in the corner. I got the contractor, Arthur Tutton, to make a tombstone at the same time that he made the one for Matt Wilson, and he put this plaque on it.

I had some rails made up too, and one day I packed across from Humbert with a bag of cement to set them up. Halfway between Bullita and Humbert the damned packhorse got away with the brumbies. A brumby stallion got in the mob that night and chased her around, and she broke the hobbles and bang! Away she went. This meant that I couldn't do anything more — the bag of cement is still lying there in the middle of the bush.

I went back to Humbert then and took the cement tombstone over to Bullita in a truck and left it there. I reckoned someday I'd take it up the Baines River and put it on Crisp's grave. This would have been a hell of a hard thing to do because it's well timbered country, with logs and rocks and anthills hidden in long grass. Well before I ever got round to it I'd sold out.

Some years ago the Northern Territory Conservation mob acquired Bullita and in 1987 they wrote to me to ask if I thought I could relocate Crisp's grave. I told them I thought I could, but it was about twenty-three years since I'd last seen it. They said, "Well there's nobody else knows where it is." I went up there and they picked me up in Katherine and we went out to Bullita. Actually Reg Durack had a bit of an idea too, and they got him in.

We had a bit of a yarn about how we were going to find it and I said to them, "There's eleven helicopters at VRD and I understand they're for charter. If you could get one over here we'd be able to pick up the grave a lot easier that way."

They got a helicopter in and the next morning we flew up the river. When we got near the place I pulled the helicopter up and said, "It was somewhere around here."

We flew around and around, and then landed. We had a good walk around looking for the grave, but the helicopter pilot was on the move to go. "Are you sure it's the right place?" he said.

"I'm nearly certain it is," I told him.

"Are you sure it's not up further again?"

"No, it's around here somewhere."

"I think we'll fly up further."

We flew up about another ten or twelve miles, just took our time and had a good look at the river and cattle. Then we came back and I said to the pilot, "If you can steady the helicopter down to about four or five miles an hour, that'll give us time to study the country. When you're flying too quick you're likely to go over the darn thing."

We flew round and the pilot was on the outside driving, and bugger me if he didn't pick the grave up! He said, "I can see three stones together and I'm nearly sure it's a grave." I was jammed in the middle and we had to circle about three times before I got a glimpse of it. We brought the helicopter down within about 100 yards of it, walked over and had a look. It was the grave all right. One or two of the rails were still there and it amazed me to see that there was still some wire on them too.

They never took that plaque from Bullita to Crisp's grave. They left it there. I said at the time, "Well I'd like to have taken it up."

"It's up to you, but we'd rather leave it here. It's going to be too rough for anyone to get up there. We'd rather it stayed here at Bullita."

I'm still not too happy about it you know. That's where Crisp was speared, that's where he was buried, and to hell with the visitors if they couldn't get up there.[4]

It's amazing you know, when I think back — talk about coincidence. I was in Mount Carmel College when my father wrote and told me about Crisp being speared. Crisp's belongings were sold and Billy Schultz bought his revolver, one of the old-time revolvers. Well eventually that came in to Woodhouse and my father had it in his menagerie case. It was there for years and years. Then when Billy Schultz got killed and we were going out there, Dad said, "Oh, we'll take this with us. It could be handy." We looked around and got cartridges for it, and I always used to carry it around. It was a little light revolver, very handy, and I always carried it on my belt. By God it was deadly, and could I shoot with the damn thing. Anything within forty or fifty yards I'd wallop every time — I didn't seem to be able to miss! I'd go up to the yard when I was shooting a killer and wait my opportunity, and I'd hit it in the forehead every time and bring it down.

Every one out in the Victoria River country carried a revolver — even the damn old cook in the mustering camp — but they had those heavy cumbersome colts and they'd fasten them onto their saddle. I could never see the sense in that; time and again when the mustering was on, somebody would end up getting a buster off a horse. *Bang* — away goes your horse with your revolver strapped onto the saddle. If you didn't catch him then he'd run under a tree or something, and he could knock it off the saddle.

I suppose I carried that revolver around over thousands of miles — into Queensland when I took cattle and whenever I was out mustering at the Humbert. It didn't matter where I went, I always had it on me. One afternoon I took a ride up the East Baines to see what cattle were around, and when I came back I ran onto a young bull about two years old. He took fright and ran down the way I was going, towards my camp. I thought, "Gee, I might end up catching this bloke and throwing him if he keeps going." I chased him down for a good half-mile or so and twice I jumped off my horse, but the beggar whipped around each time to have a look at me. Anyway, the third time I got him. I castrated him, earmarked him, and let him go.

Well when I got into the mustering camp I discovered my revolver was gone, the pouch and everything. I remembered that where I'd put my belt through the keeper, the stitching had broken away a bit — damn me, when I was running around after this bull, it must have come apart. We went back next day and had a good look around, but we had to search the best part of a half a mile, or three parts of a mile. There was a hell of a lot of grass there which had fallen over and been trampled down, and I'm damned if I could follow my tracks.

But to think that I had to come all the way back to within three miles of Crisp's grave and lose that revolver there! It had been into Queensland and on Woodhouse, and it had lain in Dad's menagerie case for four or five years before we went out to Humbert. Amazing, isn't it? It's still lying out there on the flat somewhere, I suppose. By gee I took that hard, losing that revolver.

About fifteen years after I got Bullita I got another block to the south of Humbert, the Whitewater block. In about 1960 or '61 Dan McGuinness from the Lands Department in Darwin paid me a visit and wanted a look around the place. I went in the truck with him and going along I said, "I wouldn't have minded a portion of that block across the Wickham River," and he pulled up and said, "Where do you mean?"

I said, "Across over there, even if it was only big enough for a good sized horse paddock."

We were on top of a ridge and he had a good look around, and he said, "Well I think I can do better than a horse paddock."

"How do you mean?"

"I think I can recommend that you get around 400 square miles of country. VRD has to lose 500 square miles before their leases are renewed. We've taken a 100-square mile portion and given it to the Department of Primary Production," (that's Kidman Springs), "and VRD still has to get rid of another 400 square miles. You leave it to me."

Later on he called me up to Darwin and said, "Yes, you can get 400 square miles." I couldn't believe it! It was just handed to me on a platter. Dan said, "You might be able to run about 2000 or 3000 head of cattle on it." I didn't say anything or argue the point, but I thought, "A fat lot you know."

> In the Darwin Supreme Court the proprietors of Victoria River Downs Station obtained an injunction to prevent Humbert River Station owner Chas. Schultz from mustering a block resumed from them last year and granted to him. Chas. Schultz had notified V.R.D. that he intended to take possession of all cleanskin cattle on his property on July 1st. V.R.D. claimed that they had left the cattle there when the country was resumed from them and apparently hadn't been able to find time nor men to clean them up in the intervening twelve months. The matter was settled out of court.
>
> Hoofs and Horns, September 1963:20

The Whitewater block was signed over to me in 1962, a hell of a dry year. I'll never forget it. VRD was owned by L.J. Hooker then and George Lewis was the manager. In about June or July he said, "You take delivery of it in September, and so far as we're concerned you can go onto that block anytime you like, Charlie." I told VRD I'd begin mustering the block as soon as the lease was formally transferred, but they turned round and slapped a restraining order on me! They reckoned they'd had hold-ups that stopped them from mustering the block in time for the handover of the lease.

VRD had had mustering camps in there and goodness knows what, but as usual, the way they went about things they couldn't clean it up. That suited me fine of course, but they thought they were going to bluff me and stop me from mustering there. Anyway, it was agreed right there and then that I could carry on, provided I held any VRD cattle for them. I was only too pleased to hold them. I wanted to get them off the bloody place because there were too many.

As soon as they got to hell out of the road, I got Harry McCullogh to put up about twelve miles of fencing to divide the Whitewater block from the Wickham River.

That was a good set-up for a trap because there were still a lot of cattle on the McCullogh's block. They'd water in the river and then go out again. Once the fence went in I left the gates open, and cattle were still going in onto the river and watering. At first a lot were getting caught on the fence, but they walked around and found out where the gates were. In the meantime I whipped up a makeshift yard and about a month after the fence went in I decided to trap. I wanted to get all the VRD cattle off the block.

The cattle were used to going through the fence at certain places, so I put bayonet traps at each place. The cattle walked through the bayonets on their way into the river, got a bellyfully of water, and when they came back they just hit the fence and were trapped on that side. It needed a big muster to run them down, so I put in a small paddock that ran to the makeshift yard I'd put up. I got about 1500 or 1600 head of cattle out of that, and about 700 or 800 were branders. I got any VRD bullocks away first. They were a damn nuisance, wild as hawks, so I notified VRD as to when I'd have their bullocks in the yard and I told them I wanted them shifted straight away. Well, I didn't have to tell them twice because they were beautiful big bullocks, never in hand since they were calves.

When the surveyors went through and marked out the new boundary between VRD and the Whitewater block, I saw they'd gone within about a mile of Gordon Creek and then cut straight across by Gregory's Remarkable Rock[5] — they'd cut me off from the main waters in Gordon Creek. I'd heard that they couldn't cut you off main waters like that, and as it happened the surveyor was still there. He'd just about finished the surveying, but going up in the hills he broke an axle which held him up for a fortnight. He was camped in on the river at VRD, near the racecourse, so I went and had a yarn with him.

He asked me, "Didn't VRD discuss this boundary line with you?"

I said, "They've never been near me."

"But they were supposed to work things out with you, then come back and tell me, and I'd run the line through."

"They never bloody well came near me at all!"

George Lewis had made out to the surveyor that he'd spoken with me and that I'd agreed to the boundary line.

"Well look, I'm supposed to run the line through there satisfactory to both parties, so if that line's not satisfactory, the best thing for you to do is talk to the Administrator."

I took a trip down to Darwin and explained to the Administrator what had happened. He said, "They can't do that sort of thing!"

"Well they've damn well done it."

"Look. Point out to me on the map just what they did."

I pointed it out to him and said, "There's water there at Police Hole, you see. I don't want the lot. If I could only get half of it that'd be enough."

"Where's your nearest water if you don't get that?"

"Back at the Wickham River, eighteen to twenty miles away."

"Oh, they can't do that! We'll fix that up."

Straight away he got in touch with McGuinness and sent him from Darwin to have a look at where the boundary line was running. We went out together and I pointed out just what I wanted done. McGuiness said, "Right. We can fix that." So they turned round then and organised a meeting with myself, the surveyor, and VRD.

When they sent word for me to come I went across to the waterhole and met Pat Shaw, George Lewis, and the big shot from Sydney, the general manager over all of L.J. Hooker's stations. They got me to show them where I wanted the boundary fence to go across the water. Well, I'd been there mobs and mobs of times over thirty years, and knew that Police Hole dried right back to the eastern end. Usually there's water in it for about three parts of a mile, but late in the year it dries up and leaves a bit of a spring there right at the crossing.

VRD said, "Where do you want this to go across?"

"Right on the middle there. That gives you half the

water and me half, but you won't be short of water because you've got Whip Spring just a mile or two away."

George Lewis asked, "How many cattle do you reckon come in there at Johnson's Billabong?" — That's a billabong back on the Wickham River, miles from Gordon Creek.

There were only about 700 or 800 watering at Johnson's, but I said, "The best part of 1000 or 1500."

George said, "If there's that many the buggers must have wings and fly in and out to water, because I can't see the tracks." He was trying to make out that for one thing, I had very few cattle, and for another, there were no cattle going in to Police Hole from my new block. He reckoned they were all going in to Johnson's Billabong for water. Oh, he was an old bugger, old George you know, the way he thought it was coming out of his own pocket all the time.

Well, they fixed it up so that a corner of the boundary went across Gordon Creek, and they surveyed from there straight over to Gregory's Remarkable Rock and right back to the corner near the Five Mile Creek. VRD fenced round the boundary, which suited me right to the ground because I just ran other fences straight onto it. When VRD was putting the fence up, the fencers only got out about a mile on top of the sandstone range when they broke an axle, so they said, "Oh, bugger this," and left it. That suited me too, because in the wet season cattle would get up on top of the hills, come down the Humbert side, and they automatically trapped themselves, you see.

When I got the Whitewater block I whacked up two or three paddocks there for a start, and in about three years I had nine paddocks. I had 2500 head of cattle in one paddock there all year round. All together I had about 5500 to 6000 head of cattle on that particular block.

I knew water was an urgent matter because it was a dry block. Straight away I hunted around and hired Martin Lind, and the first bore he put down on the place got water.

That's the one where the New Humbert or McCulloghs' homestead is situated now. It only gave 600 or 700 gallons an hour, but that will still water a good mob of cattle.

To encourage development of the stations, the government used to pay for any dud bores. The drillers would go down about 300 feet and if the water flow was less than 500 gallons an hour, they'd pull out and the bore was classed as a dud. If you got a flow of 500 gallons or more, you had to pay for it yourself. Every year I put down about six bores, and out of them I was lucky if I got one or two waters. I put down about forty bores altogether and I think I only got water out of about ten or twelve.

One time over on the Whitewater block when I was picking the site for a bore, Hessie and I camped for the night in a watercourse. Just about sundown I said, "Jesus, look at those owls!" I never saw so many owls in all my bloody life! There must have been 300 or 400 or 500, flying around everywhere. Just where the hell did they come from? They were such ugly looking brutes too you know, and they were only about ten or fifteen feet overhead. They weren't singing out, just flying around past our camp, and by Christ, did we have the bloody wind up! We didn't know whether they were going to land on our swag and bite us, or what.

One of my best bores was Basalt bore, put down by Gintey Gorey. It was near there where I first saw rubber bush. Several patches of rubber bush turned up on top of the range there in about 1965, and it just rocked me when I saw them. How did it get up there? The first three or four or half dozen bushes I saw, I immediately knocked down with an axe, took home and burnt, but that didn't stop it. It kept spreading, although it died out to a certain extent later on. Cattle do eat rubber bush, so that's a small consolation.

I never took chances with water. Without water you're buggered. You can have good country, but it's no good if you haven't got water. I never let more than about 1500 head of cattle get on a bore, but there are times that your feed'll cut out elsewhere and you have to shift anything up to 1000 head of cattle onto another bore that still has plenty of feed. There could already be 1000 head of cattle watering there, and in the end they might be walking out about four or five miles to get a good feed.

If cattle have to go out that far they don't usually come back to water every day — it's too far for them. They come back about every second day. Time and again I've been riding around ten or twelve miles out from a bore and I've seen cattle around there, and then of a night time I've been camped at the bore and heard them coming in to water. And that last half-mile, they're that thirsty they'll start trotting in, all stringing out, and they make straight for the trough to gorge themselves.

It gets very hot in the Victoria country. By afternoon water in the troughs can get too hot for the cattle to drink. I wanted to do something better than VRD, so I put roofs over my troughs to help the cattle get a good cool drink. The roofs were about six feet six high and nine feet wide, and I made them all steel to beat the white ants.

Another improvement I put in was an eight-foot strip of concrete down each side of the trough. Because of the concentration of cattle coming in to water, the earth gets worn away along each side. Number 3 bore on the Barkly was particularly bad like that. When I went to water my bullocks there they had to drink with their heads held high, and bloody calves were standing on their back legs to get a drink. No one else had put roofs over their troughs or concrete alongside, but later on I noticed they followed my example.

So that's how I got that particular block. Each year I got contractors to put up a paddock. I was just clearing the place and didn't like going into debt again, but I had to do that a couple of times, buying wire and windmills, and paying for the bores. The first mob of cattle I got off in

March or April each year always cleared the previous season's debt.

Looking back, I can see that I was lucky. That Whitewater block really made Humbert. A lot of it was top-class black soil country, and by the time I'd finished it was well watered, and well fenced too. I worked Whitewater for nine years until I made the big decision in 1971 and sold Humbert River.

15 Beyond Humbert

It appears almost certain now that the pastoral industry will be losing yet another family unit with the sale of Humbert River Station ... The Humbert, owned by Charlie Schultz, is probably the best improved privately owned property in the north ... We have seen the end of an era when battling pioneers like the Tom Quiltys, Charlie Schultzes and Bill Crowsons could go out into the bush with a packhorse and swag and carve a kingdom.

Hoofs & Horns, November 1971:7

My reason for selling Humbert River Station was that I got three busters in the one year, three heavy busters, and at much the same time they'd given the Aborigines full citizen rights. I said to my wife, "I think it's time we got out," because we realised the blacks couldn't take grog.

It was the last buster that put me out of commission for about three or four months. I was short of cash and I had to get some weaners away. They were being sent all the way to New South Wales and I was getting a pretty good price for them. We were mustering Ivnors Pocket in the middle of the afternoon, around about three o'clock, and we rounded up quite a mob. There was about 400 or 500 head, and in amongst them was a micky bull about twelve months old that kept on breaking out. The boys kept pushing him back into the mob and then he came round my way. I didn't want to let him go so I pushed him back again. I had to take them up to a yard to draft off the young weaners, and I thought, "I'll take him down and quieten him." I was riding a horse called Goanna, a big fierce horse that would grab the bit — and by Jesus, he'd make you hang.

This micky bull broke out near me again and I put my

horse straight at him and pushed him into the mob. He broke out somewhere else and came out near me again, so I "after" him. Next minute I heard *bang*, and I realised the girth had broken. I felt the saddle slipping up Goanna's neck and I knew he was going to drop his head and buck. Sure enough, down went his head and he threw me about six or eight feet straight ahead. I landed on the ground on my knees and rolled, and I knew that with the pace he had up Goanna was going to come straight over the top of me. That's just what happened, and as he rolled over me he stood on my thumb — this was a shod horse.

I got up and looked at my thumb. It was turned clean around and the bone was sticking out. Two or three of the boys came around to me and I sat there for quite a while. They caught the horse and fixed the saddle up for me to ride, but I said, "No I'll walk into camp" — I was only about a mile out. My head was spinning and they kept on saying, "Are you all right? Are you all right?" I said, "Yes, I'll be all right, but keep your eye on me in case I want to sit down and have a spell." Little did I realise I was suffering from concussion, and pretty badly knocked about.

I knew where I was going, but I lost perception of distance and the camp seemed to be further and further away all the time. As I was coming down, the head stockman's wife, Eve Parten, saw me leading the horse, and realised something was wrong. She walked down towards me and said, "What happened?"

"I got a bloody buster and the old head's spinning. How far to the camp?"

"That's the camp there about 200 yards off. Are you crook in the head?"

"I'm not feeling too happy."

"I'll go up and bring the car down."

I had a Landcruiser in the camp, but I said, "Ah no, I'll be right. I can walk."

I said to Paul Quigley, the jackaroo lad I had there, "I want you to drive me down to VRD. There's a trained

Sister there and she might be able to fix my thumb." We got over onto the road on the Wickham River and went in to VRD. When we pulled up I suddenly discovered I couldn't walk. When I looked at my knees they were up like footballs. God Almighty, weren't they two beauties? I said to young Quigley, "Listen. You'll have to drive up in front of the homestead." That's something no one ever did — or I never did, anyway.

I showed him where the gate was and he came round and pulled up. Ian Michael was the manager and I said to Paul, "Don't sing out. Just call Ian to one side and tell him 'Charlie's had a buster and broken a bone in his thumb, and he wants the sister to have a look at it'. And tell him I insist that we're going back to the camp tonight."

So Paul went up there and told Ian Michael, and Ian came down and said, "Been in an accident, Chas?"

"Yes Ian, and I bloody need you to give me a hit."

He looked at me and said, "Did you get a bump on the head? What's this swelling and bump you've got there?"

"That's nothing Ian, that's nothing. Look, get this bloody thing done will yer? I want to get back to Humbert."

"Are you all right?"

"Yes, of course I'm all right. Just get this done with the Sister will yer?"

With that he said, "Well, all right. Now Paul, I want you to drive up in front of the hospital while I get the Sister."

Paul drove around and I couldn't walk so they carried me in on a stretcher. The Sister came and had a look at me, and saw the bone sticking out of my thumb, and my knees swollen right up. She didn't say much. She just covered me up and went out, and was away about an hour. All the time I was wondering where the hell she'd got to. I didn't know at the time, but she'd gone across and radioed the Flying Doctor.

When she came back she said, "Charlie, the Flying Doctor will be here at seven o'clock in the morning and you're going to Darwin. You're suffering from concussion."

I remember saying, "Am I as bad as all that?" I didn't argue the point then you see.

"Yes, but nothing serious."

"Hessie's over there on her own, back at the station. She doesn't know this has happened."

"You leave that to us. We'll have all that organised."

In the morning the flying doctor picked me up and landed me in Darwin, and who should greet me at the airport but Betty and Donna. Betty was a nurse at the Darwin hospital. She walked up and had a look, and said, "How's your leg?"

I told her, and she said, "How's your arms?"

I said, "Oh, all right." Then she looked at Donna and she burst out laughing.

God All bloody mighty! I couldn't stand this any longer. "What the bloody hell are you laughin' at? Get to bloody hell out of this plane. This is no laughin' matter." Betty shook her head and said, "Dad, you want to hear the rumours that are going around here. One rumour is that a stallion savaged you. Another rumour is that you broke your back or both legs, and you've got a broken arm." Apparently it came over the morning news: "Schultz from Humbert River Station had a buster and bones broken", or some bloody thing, and everyone had been ringing her up.

Well of course, Betty went with me straight into the surgery. They knocked me out and Betty was telling me later how they were pulling pieces of grass out of my thumb, and bits of straw and mud. She said, "We thought we'd never get it cleaned." They put me in a plaster and by Christ I was in pain then. I couldn't turn, I couldn't do anything you know. Then I started to go black. I went black all over my legs and right up to my hips.

After I'd been in hospital for about a fortnight I said to the doctor, "What about letting me go home?"

"Do you think your wife could look after you?"

"Look, there's about nine or ten lubras there and my

wife's there. Between the bloody lot of them they should be able to."

He said, "All right. I'll think it over and let you know. First of all I'll see if I can get a plane."

He came back about eleven o'clock. "The plane will be ready for you in half an hour, so be ready." I went in a wheelchair to a taxi. At the airport I had about seventy or eighty yards to get out to the plane. I still don't know how I managed it, and they kept saying, "You shouldn't be going home yet!" All that had me worried was when I got back to Humbert, how I was going to get out of the plane.

We pulled up on the Humbert airstrip there and Hessie had the car fifty or sixty yards away. I beckoned her over and she pulled up about fifteen yards away. I beckoned her again and she said, "What? Do you want me to come closer?"

"Yes, right up here."

She took one look at me and said, "My God Charlie, you should be back in the hospital again."

I said, "I'm this damn far now. I'll have to sit it out."

The pilot got out and gave her a hand to put me in the car. It was my knees that were giving me the trouble. Back I went to the house and she was blowing me all the way about coming back too soon. Of course, by that time I realised myself that I shouldn't have come out. She said, "Now how are you going to get to the house? There's about a forty yard walk."

"Get that big lubra down there, Old Alice." (She was about eighteen stone, you know.) "Get her up here and I'll have to put one wing around her neck and one around yours, and between the two of yers, carry me down."

I never went through so much pain in all my life with that forty yards. I hit the bed and said, "Here I'll stop."

"Do you want dinner?"

"Listen, I want nothing. Let me die quietly."

Well, I was there for another three weeks before I got out of bloody bed, and it was nearly four months before I got on a horse again.

That buster was the deciding factor about selling Humbert. Hessie kept on saying to me, "Charlie, if anything every happened to you, the girls and I would never be able to carry on now that the Aboriginals have been given full citizen's rights. The girls don't think that the boys will work for them."

Betty had often taken the mustering camp out for a couple of days, but at the end they'd be jacking up and doing what they liked, not taking any notice of her.

I'd never have sold Humbert if I'd been able to run onto a decent manager. You get sick of a place when you're on it day in and day out, doing the same sort of job. I had one bloke there — he was a good man as long as he was under someone. About 1962 I went south for twelve months and for a while this bloke hit the grog which made things a bit awkward. I wired up two or three times, but couldn't find out where the hell he was. He got a certain amount of contract work done, but he had full control of the cheque book and I ended up owing the banks £40,000 — money spent on contract work. Luckily I had a mob of steers coming on, so I knew I could clear that and be back level again, but 40,000 was a lot of money to make up.

Three months after getting back from hospital I put the place on the market, and inside of a week I got eight genuine inquiries. Then I made a big mistake. When I put the place up for sale, instead of getting a damn solicitor I let my accountant sell the place. He came up from Adelaide and took over, and made a hell of a mess of selling it. Donna was on the right track and said to me once or twice, "I think we should get a solicitor," but I still had faith that everything was going to go all right. If you've got your faith in someone, you take chances. You think, "He knows what he's doing."

Well, he did all the talking, apart from odd times asking me a question. I thought I could understand what they were saying, but these blokes, they're talking away and the average bushman doesn't know the first thing about legal agreements. I'd never heard of a "cooling off period" for a start — not until it was too damn late. That accountant seemed to be more on the buyers' side than on mine. I think his one ambition was to get back to Adelaide as soon as he could to be able to brag to his mates down there, "I've just sold a million-acre block up in the Territory, a cattle station."

I thought the agreement was that cattle weren't to go off the station till the buyer had made the first payment, but my accountant made an agreement which allowed the buyer to send cattle off the station straight away. They bought the place, then sold the cattle off to raise the money for the deposit. I'm paid for it now, but it took ten years to get the last of the money. I was getting paid so much each year, but I wasn't getting any interest on the money owing. Then when I took my business out of that accountant's hands about six months after the sale, oh God, wasn't he arrogant!

Hessie had roughed it with me on Humbert River for thirty years, so when the place was sold I thought, "I'll give her a good holiday." I took her to New Zealand and ended up buying four New Zealand mares at the yearling sales. The next year I went back and bought two more, and I think Hessie bought one for herself. That was seven. I had a lot of wins with one horse called Brave Monarch — seven wins I think it was — and eighteen places, all on big city tracks. He won about $43,000 and more than paid for himself.

When I shipped the racehorses back I brought a 320-acre farm at a place called Parawa, south of Adelaide, between Yankalilla and Victor Harbour. Quite a few people have asked me why I chose to move to Adelaide. Well, when our kids went to high school in Adelaide they made their friends down there. My wife made friends down there too, when she'd go down to pick up or drop off the girls at school. Then we decided to have our holidays in Adelaide instead of going Queensland way. We always stopped a week or a fortnight there, and gradually made our own friends.

When I bought the farm I also bought a couple of hundred head of cattle. As time went on I decided to cut back on the work I had to do, so I sold 200 acres. I just kept 120 acres for myself — enough to run a few horses and about sixty or seventy head of cattle.

> Oh Christ! I am sick of the city,
> I wish that I was back
> In the N.T. open spaces,
> With my saddlehorse and pack.[1]

About two years after I went to live at Yankalilla I began to think that the biggest mistake I ever made in my life was to sell Humbert. I still don't know whether I did the right or wrong thing in selling the place you know, if you spend the best part of your life on a place, you're always wondering what's happening out there. Humbert is frightfully run down now. They just let it go to pieces. Buildings want painting, yards want fixing, buildings that I had there they pulled down and put up elsewhere.

Hessie passed away on June 25th, 1979. I flew up to Ayr and she was buried in her home town beside her mother and father, and the infant that was the cause of her mother's death. Hessie was only fourteen when she lost her mother; she had to look after her sister Muriel, who was only eleven years old.

I lived at Yankalilla until 1992 and had a lovely life there too, just quietly. At the last there I had five cats, about twenty-five chooks, a few mares, and a stallion. I don't know why I kept the chooks because I gave away all the eggs they laid. A fair number of visitors used to come there.

It was like on the station — I might get a run of them. You wouldn't see anyone for about a week or a fortnight, then all of a sudden everyone seemed to come at once. These days I'm living on Betty's farm, near Proserpine. I've got my own humpy there, and I can watch Betty's Brahman cattle wander around the place.

People often ask me, if I had my time over would I do it all again. I always answer, "Well there's no such thing as doing it again." I had forty-four years "Beyond the Big Run", on Humbert River station. Some years were damned hard, and lonely, but most were good. I suppose looking back now I can say that I enjoyed most of them, and don't regret any.

Notes

Chapter 1

1. James Simpson Love dominated the Indian Remount trade from the 1890s until his death in 1933. For details on Love, and on the Indian Remount trade more generally, see A.T. Yarwood, 1989. *Walers: Australian Horses Abroad*. Melbourne University Press, Carlton.
2. The *Sydney Morning Herald* of 19 December 1925 reported that Constable James Mathews had been found seriously injured in Glebe. I could not locate further information.

Chapter 2

1. *The Northern Territory Times* of 7 July 1925 reported that W.J. Schultz had met with an accident in the cattle camp and was in the "Wimmera Home Hospital" at VRD.
2. The Wimmera Nursing Home at Victoria River Downs was a hospital established by the Australian Inland Mission (Presbyterian Church). It operated between 1922 and 1939 J. Makin, *The Big Run: The Story of Victoria River Downs*. Weldon Publishing, Sydney, 1992. pp. 123–24.
3. Richard Townshend was manager of Victoria River Downs from 1904 to 1919 (Makin, p. 183).
4. Some of the stories that Charlie remembers Ivor Hall telling him were documented in an interview with Ivor's brother Noel, published in the *Northern Territory Newsletter*, January/April 1975, pp. 33–35.

Chapter 3

1. Matt Wilson died towards the end of January, 1931 (*The Northern Standard*, 3 February 1931)
2. Arthur Cahill was thrown from his horse at Weaner Yard on 23 December 1930 and died on 26 December (Humbert River Station diary).
3. The Wave Hill Police Journal records that the "Schultz Expedition"

arrived at Wave Hill on 30 September 1932, and left for Tanami on 2 October 1932 (Northern Territory Archives Service, F292).

Chapter 4

1. Ward was refused a renewal of his block, permit 56, in 1908, because the South Australian Lands Department decided to declare it an Aboriginal Reserve. Ward protested and stayed living on the block until his murder late in 1909 or early in 1910. For a detailed account of Brigalow Bill, see D. Rose. *Hidden Histories*. Aboriginal Studies Press, Canberra, 1991. pp. 119–29.
2. The Timber Creek Police Journal of 26 June 1910 records that, "Gordon was buried by the MC and Henry Bening on Light Creek on 12th inst (Northern Territory Archives Service, Darwin. F302).
3. Victoria River district Aborigines will indicate an object, place or direction by turning their face in that direction, raising their chins and projecting their lips forward (pers obs, DL).
4. Aboriginal English uses the term "him" to refer to both male and female.
5. *The Northern Standard*, 1-6-34, reported that, "Mounted Constable T. Hemmings has resigned his position in the Northern Territory Police force to take over a farming proposition in the south-west of South Australia. Tom first came to the Territory as an operator in the local telegraph office and was noted for his courtesy and efficiency".

Chapter 5

1. In 1965 a CSIRO employee, W. Arndt, stated that: "… one man, viz. the manager of Victoria River Downs (V.R.D.) station controlled 18,000 square miles, and it was extremely difficult to travel or survive in this area without his approval and material assistance" (*Oceania*, 35(4): 243).
2. According to Jock Makin in *The Big Run: The Story of Victoria River Downs*, (1992, p. 183) Hartley Magnussen was manager at VRD from 1945 to 1950. Other records suggest he was manager from 1944 to 1953. Makin also provides dates for the other VRD managers mentioned by Charlie.
3. There was a race meeting held on 26 and 27 December 1936 (Timber Creek Police Journal. Northern Territory Archives Service, Darwin, F302), and another on 27 December 1937 (Sister Joyce Falconbridge's Diary, Northern Territory Archives Service, Darwin, 853/P1).
4. The last race meeting at VRD was held in 1963 (Hooker Pastoral Company Pty. Ltd, Records: Station Report, 1959–68; Australian

National University Archive of Business and Labour, Canberra, 119/15).
5. A version of this poem, titled "Depot Races 1929", was published in *The Northern Standard*, 4-9-34.

Chapter 6

1. An angle-iron and galvanised iron homestead was erected at Gordon Creek by October 1936 (Investigation Committee. Schedule A, sheet 3. VRD).
2. Spencer left VRD in 1951 after working there for 26 years (*Hoofs and Horns*, February 1951, p. 37).
3. Texas Jack (J. Miller) and Blue Bob (R. Ballantyne) passed through Wave Hill on 14 July, 1926 (Wave Hill Police Journal, Northern Territory Archives Service, Darwin. F292).
4. Greasy Bill was reported as being very ill at VRD in October 1935 [when Charlie was on his second droving trip to Queensland]. Timber Creek Police Journal, 14-10-35. Northern Territory Archives Service, Darwin, F302.
5. *The Northern Standard* and the Timber Creek Police Journal both indicate that the tracker was named George, not Kelly.
6. Humbert Tommy committed suicide on 28-2-65. *Report of the District Welfare Officer*, M. Ivory, 7-10-65. Cited in D. Rose, 1991 *Hidden Histories*, pp. 223–24. Aboriginal Studies Press, Canberra.

Chapter 7

1. According to J. Kelly (1971. *Beef in Northern Australia*, p. 116, Australian National University Press, Canberra), the export of meat to the USA began in 1959.
2. This block was acquired by Charlie on June 16th, 1949. Northern Territory Archives Service, Darwin. F28, Box 43, GL 1335.

Chapter 8

1. These are likely to have been flocks of Little Crows (*Corvus bennetti*). This species is known to travel in large mobs and to exhibit behaviour such as that described here, particularly the "spiralling dive" which is a flock display (Ian Rowley, personal comment).
2. Government records contain numerous references to such wastage of water on the Murranji. For example, in May 1941, D.D. Smith reported that about 50,000 gallons had been wasted at Number 9 bore (Australian Archives, Darwin, CRS, F1 1940/508).
3. H.S. Hitchcock and Keith Anderson were two aviators who

perished south of the Murranji in 1929 while searching for Kingsford-Smith. For a detailed account, see P. Davis. *Kookaburra. The most compelling story in Australasia's aviation history.* (Research by Dick Smith, story by Pedr Davis). Lansdowne Press, Sydney, 1980.

Chapter 9

1. The accommodation and facilities described by Charlie are documented in *Report on Humbert River Station*, by Patrol Officer Ted Evans, 4-8-50. Australian Archives, Darwin, CRS F1 52/753 "Humbert River Station — General Correspondence".
2. Other sources suggest that VRD was not so generous. For example, see Berndt and Berndt, *End of an Era: Aboriginal Labour in the Northern Territory*, Australian Institute of Aboriginal Studies, Canberra, 1987.

Chapter 10

1. In a letter to the Lands Department, Darwin, 12-2-29, Charlie mentions that half a ton of iron had been used in construction of a house. (Northern Territory Archives Service, Darwin. F28 Box 4, GL109).

Chapter 11

1. The first aeroplane at VRD was flown by Captain Jones and arrived in 1924. The aeroplane Charlie saw was the second to visit VRD. It was carrying the Minister for Home Affairs, Mr C. Abbott, who was touring north and central Australia. The pilot was Captain Holden. Australian Archives, ACT. A1/1 Item 1938/4979.
2. According to A.S. Bingle, Vesteys General Manager, in *This Is Our Country* (privately published by his widow, Sydney, 1978), the price was fixed by the Federal Government.
3. The North Australian Observers Unit established a coast-watching camp at Blunder Bay in October, 1942. Timber Creek Police Journal, 6-10-42. Northern Territory Archives Service, Darwin. F302.

Chapter 13

1. When interviewed by F.H. Bauer in 1957, Charlie Schultz put this event as occurring in about 1948. Copy of interview in possession of DL.
2. In 1951, Ian Grabowsky was the Manager of Planning for Australian National Airways. E. Lewis, 1951. *The Story of The Air*

Beef Project in North West Australia, p. 10. F.H. Johnston Publishing Company Pty. Ltd., Sydney.
3. The "New Zealand lass" was Helen Swinburn. *Hoofs and Horns*, August 1951: 45.
4. This was the first beef cattle show to be held in the NT. It was held at Brunette Downs on 23-6-53. *Hoofs and Horns*, June 1953: 55.
5. Colonel Rose was then the Chief Veterinary Officer of the Animal Industry Division, forerunner to the present NT Department of Primary Production and Fisheries.
6. Kurt Johannsen began experimenting with road trains in 1946. J. Maddock, 1988 *A History of Road Trains in the Northern Territory 1934–88*, p. 115. Kangaroo Press, Kenthurst.
7. According to Makin p. 145, the Katherine meatworks were opened in 1963. The road from Katherine to Western Australia (the Victoria Highway) was sealed in the late 1960s. *The Inland Review* Vol. 1, No. 4: 23-27.

Chapter 14

1. Charlie Schultz had acquired the Bullita block by December, 1947. Letter from CS to A.M. Blain (MHR), 26-12-47. Northern Territory Archives Service, Darwin. F28, Box 43, GL 1335.
2. These blocks were resumed from VRD in 1952. *The Northern Territory News*, 26-5-52.
3. Dr Fred Ryle was with the Royal Flying Doctor Service at Wyndham from 1940 until 1961. M. Page, 1977 *The Flying Doctor Story 1928–78*, p. 166. Rigby Ltd, Sydney.
4. In 1991, the Conservation Commission of the Northern Territory placed the tombstone on Crisp's grave and enclosed it with a rail.
5. Gregory's Remarkable Pillar, a sandstone outcrop named to honour explorer A.C. Gregory who came across it in 1856.

Chapter 15

1. From 'Nostalgia', a poem by W. Miller (alias Billy Linklatter), nd. Copy in possession of DL.

Key Events in Humbert River History

1856　A.C. Gregory's expedition travels up Wickham River and passes the junction of Wickham and Humbert Rivers on 15 January 1856. Gregory does not name the Humbert.

1879　Alexander Forrest's expedition enters headwaters of a new river, names it the Humbert, travels down to junction with Wickham River, passes near present Humbert River homestead site on August 14th, 1879.

1883　Victoria River Downs and Wave Hill stations first stocked.

1904　W.J.J. Ward (Brigalow Bill) obtains annual grazing licence (permit 56) for 579 square miles on Humbert River, becomes first white settler on the Humbert.

1908　Decision made to use Ward's block for an Aboriginal reserve. Ward protests and remains on the block.

1909　The "Mudbura Aboriginal Reserve" proclaimed; Ward speared and killed at his homestead.

1910　Several Aborigines arrested for Ward's murder, "Gordon" shot dead.

1914　William (Billy) Butler obtains grazing licence over the so-called Mudbura Aboriginal Reserve.

1919　Butler sells rights and improvements to Charlie Schultz's father, C.F. Schultz. Billy Schultz takes over as manager. Jim Crisp speared on Bullita station.

1927　Billy Schultz thrown from a horse and killed near VRD homestead.

Key Events

1928	Charlie Schultz and his father arrive at Humbert River in January. Charlie begins life alone on Humbert in October.
1930	Fred Mork takes Humbert cattle to Alice Springs, loses 250 head.
1932	Arthur Grace loses several hundred head of Humbert cattle on way to Alice Springs.
1933	Roley Bowery loses several hundred head on way to Alice Springs.
1935	Charlie begins first of four bi-annual droving trips to Queensland.
1941	Hessie Graham and Charlie married in April. Debt on Humbert River cleared.
1947	Bullita station taken up by Charlie.
1948	Charlie first to transport livestock by truck in the Northern Territory.
1949	"Up-river block" (Upper Wickham) taken up by Charlie.
1951	Charlie initiates first airlift of cattle into the Northern Territory.
1952	Killarney, Montejinni and Camfield resumed from VRD and put up for ballot. Charlie applied, unsuccessfully.
1959	Charlie first to move cattle out of the Victoria River district by truck.
1962	Charlie acquires Whitewater block.
1965	Charlie sells Bullita to Happy Berlowitz.
1967	Aborigines acquire citizenship rights.
1970	Charlie suffers serious fall from horse, decides to sell Humbert River.
1971	Humbert River sold to Charlie Clark from Queensland.
1979	Hessie Schultz passed away.
1985	Northern and western portions of Humbert River, all of Bullita, and portions of several other stations combined to form the Gregory National Park.

Glossary

Bagman: Out of work bushman who travels by horse.

Bayonet-trap: Device to allow cattle to pass through a fence in one direction.

Big House: Residence of the manager or owner of a station.

Billibung: Colloquial pronunciation of billabong, a waterhole outside the main stream of a river.

Bo-yab: Colloquial pronunciation of the 'boab' or 'baobab' tree *(Adansonia gregorii)*

Bronco horse/mule: Very strong horse or mule especially trained to drag a beast to a rail for branding, castration or other treatment.

Bronco panel: Post and rails against which a beast is dragged.

Bronco yard: Yard specifically designed for bronco work.

Bully fire: An 'old man' fire — a big fire made with logs or stumps.

Cheeky: Dangerous, life-threatening, eg 'cheeky snakes', 'cheeky blacks'.

Coachers: Quiet cattle, into a mob of which wild cattle are driven during a muster.

Death adder: A loner, usually elderly.

Gilgai: Small ephemeral pool in clay soil away from a watercourse.

Hands-gallop: Galloping flat out.

Hit the top rail: React angrily, to lose ones temper.

Killer: Beast to be killed and eaten locally rather than sent to the meatworks.

Micky: Young wild or unbranded bull.

Myall: Ignorant or wild, usually in reference to Aborigines.

Open bronco: Bronco work conducted in the bush, using a tree instead of a panel.

Piker: Aged bullock.

Poddy-dodger: Cattle thief.

Scoot, on the: Drinking alcohol.

Shelly: Aged, as in shelly posts, shelly bullocks.

Stub yard: Yard made by standing posts side by side rather than using rails between posts.

Sulky: Angry, potentially dangerous.

Tailer, horse-tailer: Last in a mob; the person who rounds up and looks after horses.

Turkeys nest: Round earthen dam.

Turnout: Collection of possessions: "he had a good turnout there"; alternatively an event: "what sort of a turnout is this?"

Yellafella: Person with one parent Aboriginal and one European. The terms "yellafella", "whitefella", and "blackfella" are considered by some to be derogatory, but they are in general use among Aborigines and Europeans in the Victoria River district and other areas of the outback. Of these, only "yellafella" commonly has negative connotations.

Select Bibliography

Books

Barker, H.M. 1966. *Droving Days*. Sir Isaac Pitman & Sons Ltd, Melbourne.
Bolton, G. 1963. *A Thousand Miles Away: A History of North Queensland to 1920*. Jacaranda/ANU, Brisbane.
Broughton, G. 1965. *Turn Again Home*. The Jacaranda Press, Sydney.
Cayley, N. 1984 (1931). *What Bird Is That?* Angus & Robertson Publishers, Sydney.
Chauvel, E. 1973. *My Life with Charles Chauvel*. Shakespeare Head Press, Sydney.
Gee, L.C.E. 1926. *Bush Tracks and Gold Fields: Reminiscences of Australia's "Back of Beyond"*. F.W. Preece and Sons, Adelaide.
Hall, R.A. 1989. *The Black Diggers: Aborigines and Torres Strait Islanders in the Second World War*. Allen & Unwin. Sydney.
Rose, D. 1991. *Hidden Histories: Black Stories from Victoria River Downs, Humbert River and Wave Hill Stations*. Australian Institute of Aboriginal Studies, Canberra.
Willey, K. 1964. *Eaters of the Lotus*. The Jacaranda Press, Sydney.

Government Publications and Reports

Report of the Administrator of the Northern Territory, *Commonwealth Parliamentary Papers*, 31 July 1920.
Payne-Fletcher Report, 1937. *Report of the Board of Inquiry Appointed to Inquire into the Land and Land Industries of the Northern Territory of Australia*. Commonwealth of Australia, Canberra.

Northern Territory Archives, Darwin

Bow Hill Police Station Journal, 1918. F292.

Timber Creek Police Journal, 1910, 1937, 1936, 1941, 1942. F302.
Timber Creek Police Letter Book, 1929. F720.
Wave Hill Police Journal, 1918, 1924. F292.
Schultz, Charlie. Letter to Lands Department, Darwin, 12 February 1929. F28. Box 4, GL109.
Schultz, Charlie. Letter to Northern Territory Administrator, 8 August 1944. F28, Box 4, GL 109.
Schultz, Charlie. Letter to A.M. Blain (MHR), 26 December 1947. F28, Box 43, GL 1335.
Davidson, W.J. (Chief Clerk, Lands and Survey Branch, Darwin). Letter to Charlie Schultz, 28 August 1952. F28, Box 43, GL 1335.
Brodie, F.A. and Co. Ltd (Stock and Station Agents). Letter to the Administrator, Home and Territories Department, Melbourne, 23 April 1919. F28, Box 4, GL 109.

Australian Archives, (ACT)

Sheridan to Police Commissioner, 1926. Department of Home and Territories, Correspondence files, Annual Single Number Series, *1903–38: Victoria River N.T. Fight between Police and Natives*. CRS A1, Item 1926/2816.
W.J.J. Ward to Timber Creek Police, 28 July 1908. Department of Territories, Correspondence Dockets, CA60 NT Series *"Aboriginal Reserve — Ord River District, NT — Re Proposed"* CRS A1640, Item 1906/223.
Investigation of Pastoral Leases — Northern Territory 1934. Series A52/1, Item 333/51, p.12.
Wise, F.J.S. Confidential Report to Sir Charles Nathan, C.B.E., 15 August 1929. Series A494/1, Item 902/1/82.

Australian Archives (NT)

Evans, T. *Report on Humbert River Station*, 4 August 1950. Australian Archives, Darwin. CRS F1 52/753 "Humbert River Station — General Correspondence".
Schultz, Charlie. Letter to Mr Moy, Native Affairs, Darwin. 20 September 1948. CRS F1 52/753 "Humbert River Station — General Correspondence". E.C.E./11.
The Secretary, Department of Territories, Canberra, 25 January 1952. CRS F1, Item 52/250.

Archives of Business and Labour, Australian National University, Canberra

Report on VRD, 1959. Author unknown. Hooker Pastoral Company Pty Ltd, Records. Deposit 119/15.

Reports on VRD, 1963 and 1967, by George Lewis. Hooker Pastoral Company Pty Ltd, Records. Deposit 119/15.

Martin, A. Letter to Eicharn, 25 March 1935. Bovril Australian Estates Pty Ltd, Correspondence, 1933–37. Deposit 87/8.

Martin, A. Letter to Lord Luke, 9 March 1942. Bovril Australian Estates, Pty Ltd, Correspondence, 1939–47. Deposit 119/6.

Correspondence between Australian Mercantile Bank and Finance Co. Ltd, Sydney, and B.A.E. Ltd, London, and station managers, 1939–47. Deposit 119/6.

National Library of Australia

Stewart, Sister V. Letter to A.I.M. Head Office, 19 January 1937. Ms 5574: Australian Inland Mission Correspondence (various dates), Box 108.

Australian National Trust (N.T.)

Condon, H. to Timber Creek Police, 31 March 1910. Timber Creek Police Letter Book, 1894–1911. Copy held by the Australian National Trust, Darwin office.

Manuscripts, Diaries and Personal Papers

Humbert River Station Diaries (in possession of Charlie Schultz).

Diary of J.R.B. Love, 1912–14. Typed copy in the library of the Stockman's Hall of Fame, Brisbane office.

Timber Creek Police Letter Book, 1894–1911. Copy held at National Trust, Darwin.

C. Boulter, *Reminiscences of a Wanderer* [1913]. Unpublished manuscript. Copy held by the Northern Territory Historical Society, Darwin.

Unsigned hand-written manuscript (nd), believed written by W. Linklater. Original in possession of Mrs Bobby Buchanan, Darwin.

Bauer, F.H. Personal notes on Humbert River Station, based on an interview with Charlie Schultz in 1957. In possession of F.H. "Slim" Bauer, Canberra.

Index

Aborigines, conditions on stations 126-32
aeroplane, first at VRD 157-58
Afghans 152-53
airlifting cattle 189-93
air service (Connellan) 158
Anthony Lagoon Station 115, 118, 171
army 140, 143-44, 159, 161-65
Auvergne Station 54, 56, 201
Ayr 3, 6-7, 9, 21, 25, 41, 124, 137, 139, 143, 153, 222

Barwick, Alf 172-75
Bathern, Harry 116-18
Beetaloo Station 116-18
Blue Bob 86
bores 211-13
Boomundoo 48
Bowery, Roley 40-41, 67, 69-70, 82, 89-91, 95, 109, 111, 141, 144, 149, 156, 171
Brady, "Boomerang" Jack 113-16
Brady's Grave 113, 116
Brigalow Bill, see Ward, W.J.J.
brumbies 28-33, 204
Bruten, Morris 176-78
Bullita Station 37-38, 50, 53-57, 70, 75, 102, 105, 161, 196, 198-99, 203-205, 207
bullocks, shoeing 105, 120-21
bulls, attacks by 79-80, 200
bulls, shooting of 101
bulls, throwing of 79-80, 102-103, 206
"Bullwady Bates", see Bathern, Harry
Burdekin River 3, 5, 7
Butler, Billy 17, 21
Byers, Wason 170-75

Cahill, Arthur 40, 44

Cahill, Tom 40, 58
Calcutta 11-12, 14
Camfield Station 196-98
Campbell, Doug 86-88
camels 18, 133, 152-53, 186
Carrol, Dan 56-57
cattle rushes 106, 109-11
cattle, spearing of 42, 50-51, 58-61
Centre Camp (VRD) 71, 81
Charcoal (Aboriginal stockman) 86-88, 91
Chinese 22
Christmas sports, VRD 66-68
Cleanskin cattle 36, 79, 95-96, 101, 171, 197
"Colorada" Jack 56
Condon, Harry 54
Conway, Paddy 184-85
Coolibah Station 41, 170-71
Crisp, Jim 50-54, 60, 203-207
crocodile shooting 7-9
Cronin, Don 35, 40
Crouch, Jungari 122-23
crows 118-19, 144
Cusack Creek 99-100
Cunningham, A.W.H. 5, 8, 19, 25-26, 106, 125, 137

Daly (Aboriginal stockman) 133-35
Dashwood yard 62-63, 154
Depot, Timber Creek 37-39, 53, 70-71, 74, 152, 162, 176
Delamere Station 105, 178
dingoes 7, 31, 34, 77, 97, 122, 145
donkey teams 21, 41, 140, 152
Drew, Bert 182
drought 198-99
Durack, Reg 204

East Baines River 50, 56, 203-204, 206
Edwards, Jim 196-97

Eva Downs Station 112-13, 116-18
Farquharson, Jack 64-65
Fitzer, Constable Tas 91, 176-79
Fitzgerald, Constable 88-90
Fletcher brothers 4
floods 83, 156-57
Fogarty, Dave 41

gold prospecting 18-19, 41, 84
Gordon (Aboriginal outlaw) 44, 46-50
Gordon, Constable John 134
Gordon Creek outstation 69, 71, 76-78, 80-83, 86-88, 95, 109, 132-33, 138, 149, 161, 179
Gordon Creek stockboys 86-87
Grace, Arthur 106-109
Graham, Tom 94
Greasy Bill (Aboriginal stockman) 87

half-castes, removal by government 146-48
half-castes, raised at Humbert 148-51
Hall, Ivor 21-22, 44-46
Hall, Noel 21-22
Halls Pocket 21, 37, 97-98
Hemmings, Constable Tom 46, 56-57
helicopters 77, 197, 204
horse breaking 23-24
horse stealing 107-109, 172-75, 185
horses, falls from 20, 28-33, 40, 149-50, 181-82, 215-16
hospital, Australian Inland Mission 20, 66, 69, 162, 179
Humbert airstrip 166-68
Humbert Charlie 58, 60
Humbert, Les 149-51
Humbert, selling 220-21
Humbert Tommy 88-92
Hunt, George 74-75

India 9-16, 184
Ivnors Pocket 21, 24, 135, 149, 192, 200, 215

Jasper Gorge 162
Johns, Larry 149-51
Kaiser Bill (Aboriginal stockman) 92
Kelly, Tracker 90-92

Killarney Station 196-98
Knox, Jack 66-67, 69, 76

Lake Nash Station 118-19, 124
Lancewood yards 186-88
Larkin, Les 35
Lennox, Stan 9
Lewis, George 64, 71-72, 74, 208, 210-11
Lewis, Jack 153-55
Liddy, Tom 20
Light Creek 47-48, 60-61, 88
Linklatter, Billy (Billy Miller) 75
Long Barney, death of 133-34
Love, J.S. 9-11, 20

Macalvry, Dan 79
McPherson, Ian 11, 13
Magnussen, Hartley 63-64, 71, 191
Magoffin, Dave 41
Mahnikee 69
Mann, Len 159-60
Maroun (Aboriginal outlaw) 43-44, 60
Martin, Alf 20, 22, 25-26, 70, 81, 87-88, 94, 103, 132, 140
Martin, Mrs Alf 138, 140
McDonald, Bill 190-92
McDonald, Jack 80-81, 84
McColl, Scott 64, 201
McGuggan, Alec 84-85
McGuigan, Jack 84-85
McGuinness, Dan 207-208, 210
Michael, Ian 64, 217
milking herd 135-36
Montejinni outstation and Station 42, 62, 69, 71, 84, 92, 109, 152, 160, 173-74, 196-98
Moolooloo outstation 42, 69, 71, 76, 152, 160, 181-82, 186-87
Morck, Jack 105-106
Mount Sanford outstation 42, 69, 71, 86, 94-95, 103, 152
Murranji Track 62, 109-13, 121-24, 157, 186-87, 194

Neumayer Valley Station 18-19
Nelson, Bob 69-70, 81
Newcastle Waters Station 84, 106-107, 109-12, 115-18, 123, 153, 157, 193-95

Index

Noble, Jack 83-84
nurses, Australian Inland Mission 66-67, 72-73, 179

O'Reily, Jerry 175-79

Peter Creek 34, 47, 108-109, 133, 138, 140, 157, 167
Pigeon Hole outstation 42, 69, 71, 81-82, 103, 109, 152, 173, 176, 189, 193
Pine Creek 22, 155
Police Creek 37, 46, 96
Potts, Constable Jack 146-47

Quirk, Jack 64-65, 166-68, 194-95

Races, VRD 68-72
Reynolds, Frank 82, 87-88, 95, 133, 138
Riley, Old (Aboriginal stockman) 20, 36, 43, 60
Riley Pocket 34, 58-59
road trains, in Victoria River district 193-95
Roden, Jack 22, 68, 136, 142
Rogers, Jackie 179-82

Sailor (Aboriginal stockman) 87
Schultz, Albert 5, 17-19, 41
Schultz, Betty 144-46, 199-200, 218, 223
Schultz, Billy 16-21, 28-32, 34, 50, 76, 105, 205-206
Schultz, C.F. (Charlie's father) 3-6, 9, 17-21, 40-41, 86, 205
Schultz, Donna 142-46, 184, 199, 218, 220, 223
Schultz, Hessie 63, 73, 128-29, 131, 133-34, 137-46, 148-51, 167-68, 199-202, 212, 218-19, 221-22
Schultz, Phyllis 124, 137
Seaton, Billy 28-33

Shaw, Snowy 62, 69, 92, 173-75
Sheridan, Constable Frank 175-81
Simpson, Tom 83-84, 114, 132
Snowy (Aboriginal stockman) 134
soap making 128-29
Spencer, Frank 56, 69-70, 76, 79, 82, 87-88, 103, 109
Steep Creek 89, 157, 187, 194
stone buildings 139-40
Stone Yard 98-101, 103
Stott, Constable Gordon 90, 115-16, 171-72

Tanami Desert 18, 41
Timber Creek 36, 39, 46, 51, 53, 56, 74, 88, 148-49, 152, 162, 174, 176, 193
trucking cattle and horses 183-84
Townshend, Richard 21, 45

up-river block 103-104
Upper Wickham 92-104, 109

Vesteys 111, 152, 159

Ward, W.J.J. (Brigalow Bill) 43-45, 48, 60
Wave Hill Station 20, 40, 65, 71, 83, 84, 93, 105, 107-109, 113-15, 123, 138, 152-53, 157, 159-60, 175, 179, 184-85
Wilkins, "Peggy" 85-86
Wilson, Matt 36-39, 107, 204
Wimmera Nursing Home, *see* hospital
windmills 140-41, 213
Woodhouse Station 5-7, 9, 16, 18, 23, 25, 40, 111, 121, 124, 137, 139, 206-207
Wyndham 38, 65, 71, 77, 82, 106-107, 173, 190, 199, 201-202

Yankalilla 221-22